Springer Theses

Recognizing Outstanding Ph.D. Research

For further volumes:
http://www.springer.com/series/8790

Aims and Scope

The series "Springer Theses" brings together a selection of the very best Ph.D. theses from around the world and across the physical sciences. Nominated and endorsed by two recognized specialists, each published volume has been selected for its scientific excellence and the high impact of its contents for the pertinent field of research. For greater accessibility to non-specialists, the published versions include an extended introduction, as well as a foreword by the student's supervisor explaining the special relevance of the work for the field. As a whole, the series will provide a valuable resource both for newcomers to the research fields described, and for other scientists seeking detailed background information on special questions. Finally, it provides an accredited documentation of the valuable contributions made by today's younger generation of scientists.

Theses are accepted into the series by invited nomination only and must fulfill all of the following criteria

- They must be written in good English.
- The topic should fall within the confines of Chemistry, Physics, Earth Sciences, Engineering and related interdisciplinary fields such as Materials, Nanoscience, Chemical Engineering, Complex Systems and Biophysics.
- The work reported in the thesis must represent a significant scientific advance.
- If the thesis includes previously published material, permission to reproduce this must be gained from the respective copyright holder.
- They must have been examined and passed during the 12 months prior to nomination.
- Each thesis should include a foreword by the supervisor outlining the significance of its content.
- The theses should have a clearly defined structure including an introduction accessible to scientists not expert in that particular field.

Vincent Traag

Algorithms and Dynamical Models for Communities and Reputation in Social Networks

Doctoral Thesis accepted by
the Catholic University of Louvain, Belgium

Springer

Author
Dr. Vincent Traag
KITLV
Leiden
The Netherlands

Supervisors
Prof. Paul Van Dooren
Department of Mathematical Engineering—ICTEAM
Université catholique de Louvain
Louvain-la-Neuve
Belgium

Prof. Yurii Nesterov
Center for Operations Research and Econometrics (CORE)
Université catholique de Louvain
Louvain-la-Neuve
Belgium

ISSN 2190-5053 ISSN 2190-5061 (electronic)
ISBN 978-3-319-36062-1 ISBN 978-3-319-06391-1 (eBook)
DOI 10.1007/978-3-319-06391-1
Springer Cham Heidelberg New York Dordrecht London

Printed on acid-free paper

Springer is part of Springer Science+Business Media (www.springer.com)

Supervisors' Foreword

We are living in a world where the amount of data that is collected and stored is just staggering. Moreover, the information and communication technology required to have access to these data has become quite affordable so that everybody who wishes can have access to it, as far as it is in the public domain. This has had a tremendous impact not only in science and technology but also in commerce and recreation, where having access to the right bit of information is crucial. An obvious example of such a source of information is the "internet," with which we mean the World Wide Web and search engines such as Google. But social networks have started to play a big role as well in getting access to data. Networks such as Facebook, LinkedIn, and Twitter have attracted billions of users in a very short time. These networks allow friends or colleagues to connect to each other and retrieve or distribute information that would be hard to find otherwise. But the networks themselves can also be viewed as data that can be analyzed to extract valuable information about the "nodes" of the network, which can be people, but also objects, pictures, texts, and so on.

The structure of such networks plays an important role in the type of information one can extract from them. One prominent feature of many social networks is the clustering of nodes (people in this case). Friends tend to have many friends in common, thereby creating social groups in which many people know each other (and often have the same taste, behavior or habits). Knowing these social groups yields additional insight into the structure of these networks and can be used for commercial purposes by companies or by providers of certain services. To find these groups, the idea is to look for densely connected subgraphs in the network, which are only loosely connected among each other. These are commonly known as "communities" and the field that deals with finding such communities is known as "community detection." Several more mathematical criteria have been proposed to characterize these groups more precisely, such as the popular method called "modularity," introduced by Newman and Girvan. In this book, the author analyzes in depth the problem of community detection and proposes an alternative method, called the Constant Potts Model, and explains that its major advantage is that it has no resolution limit and hence can also detect relatively small communities in large networks. Although the proposed solution does not suffer from the resolution limit, there are still some questions related to scale. The author then introduces the concept of "significance" which helps to decide whether a partition should be rather coarse of rather fine. Both these developments are important contributions of his work.

Although most methods for community detection focus on networks that have positive links, negative links also appear naturally and may represent animosity or distrust. Incorporating these negative links can be done in a relatively natural manner by insisting on as little negative links as possible within a community. This is illustrated here using a network of international relations and a citation network. The structure of negative links has been studied by the social sciences before in the context of "social balance" and is based on the adage that "the enemy of an enemy is a friend." The main observation in that literature was that socially balanced networks can be split into at most two factions where each faction has only positive links within and negative links between the factions. Besides the important question of detecting such factions in networks, the author also analyzes how social balance may emerge and why it is observed so often. This is done using a new dynamical model that explains the emergence of social balance. In addition, there is a natural connection between negative links and the problem of the evolution of cooperation that one finds in the area of dynamical games. The author uses ideas borrowed from this literature to explain that social balance can lead to cooperation. Finally, the author also looks at how to determine who will cooperate with whom. This is especially pertinent in online markets such as eBay or Amazon, where one wants to make sure one can trust ones "friends." The author shows how to use the network consisting of local links (which are positive for "trust" and negative for "distrust") to calculate a global trust value, which is the "reputation" of the corresponding node.

This book makes the bridge between two distinct areas: (i) community detection in large sparse graphs and (ii) social balance and evolution of cooperation. The author covers quite a wide range of topics in it since the two distinct areas require different backgrounds. The synthesis of the state of the art in these areas is well equilibrated and all the important concepts are well described. The book makes important novel contributions in a very competitive area of research.

Louvain-la-Neuve, April 2014

Prof. Paul Van Dooren
Prof. Yurii Nesterov

Preface

The first presentation ever of my research was on February 2009, Friday the 13th—how scary is that—and was in front of mathematicians in Louvain-la-Neuve—how scary is *that*. Having only a Master's in Sociology in my pocket I arrived there to apply for a position as a Ph.D. candidate (although, if memory serves me well, that was not entirely clear for everyone). Of course, I was no complete stranger to mathematics, yet not having studied it and still wanting to pursue a Ph.D. in that direction did not quite seem to add up. Fortunately, my advisors Paul Van Dooren and Yurii Nesterov were happy to take me on board. I am grateful to this date that they did so. The leeway they allowed me to pursue my own interest is much appreciated. I have learned a lot from them, and both are impressively (if not intimidatingly) fast when doing mathematics. I was fortunate enough to be funded by the *Actions de recherche concertées, Large Graphs and Networks* of the *Communauté Française de Belgique* and the Belgian Network Dynamical Systems, Control, and Optimization (DYSCO), funded by the Inter-university Attraction Poles Programme, initiated by the Belgian State, Science Policy Office.

My fellow Ph.D. students have also taught me a lot. Not having had the exact same training as most other Ph.D. candidates, I could borrow their expertise in trying to understand something. For some courses I was the designated teaching assistant, without actually ever having taken the course myself, making it somewhat of a challenge. For example, I had to learn integer programming. Before being able to learn integer programming, I had to learn linear programming, which also involved doing the simplex algorithm. If I say I will never forget that, it is probably true, but I would like to never make another simplex tableau again.

Around the time I started, there were a few other students coming in from the private sector: Pierre, François-Xavier, and Arnaud, which reassured me that I was not the only one that had tried the private sector and returned to academia. Throughout the years, Arnaud and I collaborated on various projects, I have enjoyed our cooperation very much. Similarly for Pierre Deville and Adeline Decuyper, it was a pleasure working with you, and good luck organizing NetMob next time around, for which Vincent Blondel was kind enough to invite us last year. Finally, I would like to thank everybody else in the Euler building (too many people to list) for the great atmosphere during coffee breaks and lunch time. I have enjoyed the conversations in the cafeteria very much, although for the most part I have only listened instead of actually engaging in the discussions.

I would like to thank the other members of the jury, François Glineur, Vincent Blondel, Marco Saerens, and Patrick De Leenheer. Their comments and remarks have greatly improved this thesis. I have had the pleasure to collaborate with Patrick while he was Belgium in 2012. His help was quintessential to the progress on the social balance project, for which I am much obliged.

Many friends and family have come to visit in Brussels, and it was always a pleasure having you. Bas, Hans-Hein, and Mathijs, you have always had that fingerspitzengefühl for coming to Brussels. Merijn, despite your busy job, two kids, moving two times, and an entire renovation, you still managed to come to Brussels: so good you could make it. Roel, our discussions on the balcony of the Rue Lebeau were marvellous—as always—I hope to continue many of them in Amsterdam. Many a Sunday morning was spent at the Vossenplein/Place du Jeu de Balle when my family-in-law came over. Fortunately, due to long breakfasts we never arrived that early, you're always welcome for such long breakfasts. From Brussels, I have very much enjoyed climbing with you Tom, I hope to see you still after moving. Frank, our lunches were a pleasant distraction from the daily Ph.D. grind. Many friends go unnamed, but not forgotten: I hope to see you all more often when I am back in Amsterdam. Likewise for my parents, my brother and sister, Ernst and Susan, I hope to see you more often, Marco, Carlijn and Niels included of course. I hold you all very dear. Mom and dad, you have always supported me—both before and during my Ph.D.—I will always be grateful for your care and love.

Finally, somebody that merits a paragraph in its own. The first two years of my Ph.D. our time together largely loomed in the shadow of the loss of your mother. Although such a loss will always leave a void, together I believe we have overcome. After having been parted by over 200 km of rail for over 3 years, we finally spent the last year together in Brussels. It was a bliss to finally live together, and I hope to continue to enjoy your company for many years to come! Lio, you are my true love.

Contents

Nomenclature

$\mathcal{C}$	Community sets
$\Delta\mathcal{H}$	Difference between two partitions
$\Delta\mathcal{H}(\sigma_i = c \mapsto d)$	Move node
$\Delta\mathcal{H}(\{c,d\} \mapsto c')$	Merge communities
$\Delta\mathcal{H}(c' \mapsto \{c,d\})$	Split communities
δ	Kronecker de, Dirac delta
$\mathcal{H}(\sigma)$	Canonical model
$\mathcal{H}_{LP}$	LP model
$\mathcal{H}_{\text{AFG}}$	AFG model
$\mathcal{H}_{\text{CPM}}$	CPM model
$\mathcal{H}_{\text{RB}}$	RB model
$\mathcal{H}_{\text{RN}}$	RN model
$\langle\cdot\rangle$	Average
μ	Mixing parameter
NMI(X, Y)	Normalized mutual information
σ	Membership vector
VI(X, Y)	Variation of information
A	Adjacency matrix
B	Modularity matrix
E	Edges
$E^{\pm}$	Positive/negative edges
F	Faction
$f_s(\vec{n})$	Fitness
G	Graph
$G^{\pm}$	Positive/negative graph
$H(X)$	Entropy
$H(X, Y)$	Joint entropy
$H(X \mid Y)$	Conditional entropy

$I(x)$	Information
$I(X, Y)$	Mutual information
I_n	Identity matrix
k_i	Degree
S	Community matrix
V	Nodes

Chapter 1
Introduction

Social networks have become increasingly more prominent the last decade. The advent of online social networks have attracted the interest of millions of people. They allow friends to connect over the internet, and share whatever they want with each other. Facebook was only launched in 2004, and has started out with a few thousand people, but currently over 1 billion people use its services. Although the online social network of competitor Google was rolled out only in 2011, they apparently have succeeded in attracting over 500 million people. Other services such as LinkedIn use a more professional career orientation and have a smaller user base of only about 90 million users. Twitter, with its well known short messages, has grown to half a billion users in only 6 years time. They handle more than 300 million tweets per day, some 3,500 messages per second.

The structure of these networks is fascinating, and gives us a glimpse of how people connect to each other. Yet thinking about social networks has a long history. Some of the oldest hypotheses, can only be studied now that data has become available in such overwhelming amounts. For example, it was suggested by Granovetter [6] that people that have many common friends have a stronger connection, an effect that was recently corroborated by Onnela [18] by using mobile phone data. Before that, it was suggested by Heider [11] that friends tend to share both friends and enemies, something that was also found by Szell et al. [24] in a network of friends and foes in a massive multiplayer online game. Similarly, Simmel [22] argued that triads in which all three people know each other should appear quite frequent, something known today as clustering. In a famous experiment, Milgram [15] analysed chains of letters sent across the US, and concluded that it took only six intermediaries on average to reach a random person in the US. This combination of the "six degrees of separation" and high clustering led Watts and Strogatz [26] to create a model of this so-called "small world". Recently, it was also confirmed at a global scale by Backstrom [2] using Facebook data, but they found that users are only four steps away from each other on average.

This thesis addresses issues in social networks and is divided in two parts. Both parts address two different broad topics, but they are not completely unrelated.

V. Traag, *Algorithms and Dynamical Models for Communities and Reputation in Social Networks*, Springer Theses, DOI: 10.1007/978-3-319-06391-1_1,

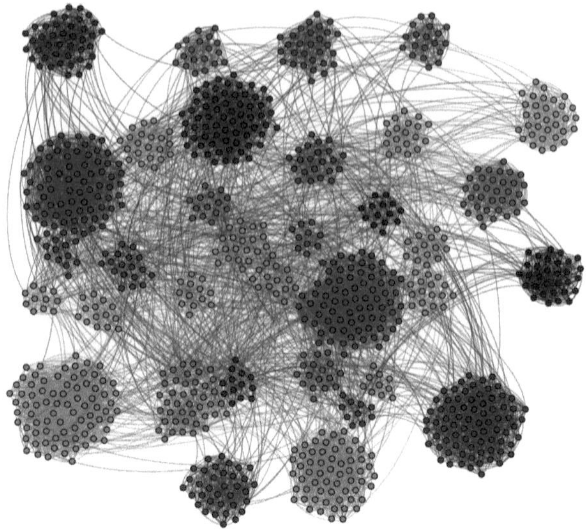

Fig. 1.1 Example of communities in networks

The first part focuses on identifying groups in social networks and in the second part we will study reputation and cooperation in networks. The first topic arises naturally because of the high clustering in social networks: if people tend to have many friends in common, they probably form some sort of a social group (Fig. 1.1). However, suppose we are only given a network, but not which people belong to what social group. Could you then still identify groups of people?

This has been one of the major challenges of the past few years. But as so many other phenomena, this subject has a rich history. Sociologists understood that many networks can be divided into groups in a meaningful manner. For example, in what is probably the most famous network, Zachary [27] gathered data on a karate club. There was a row over prices at this club, and the club split in two groups. Surprisingly, to which group people belonged could be accurately predicted on the basis of their social relationships. Another famous example revolves around monks in a monastery [21]. Some of the ongoing practices at the monastery were questioned by some novices, and the social networks could be divided in four different groups that opposed or defended these practices. But also in historical context social groups can be identified, and Padgett and Ansell [19] identified the Medici group as much more centralised than the oligarch faction in medieval Florentine politics. But also in other fields, the idea of having communities is quite natural. In networks of international trade, some countries trade much with each other, but not so much with others [3]. For example, many Western countries trade more amongst each other than with others. But also technological networks such as the world wide web contain communities: websites of related content refer mostly to each other [13]. These communities then represent common topics, such as politics, football or auto mobiles. Biological networks, such as food webs—which species eats which species—have communities in the form of ecological subsystems, a phenomenon also known as compartmentalization [23].

For example, in the ocean, many species live in the top of the ocean, hence feeding only on other species which live there, while completely different species exist at greater depths. We might also mention biochemical networks such as protein–protein interaction or metabolic networks, where communities seem to represent proteins or metabolites with similar functions [8]. Many additional examples of communities in networks could be provided [7, 9, 12, 14, 18, 20, 28].

But this subject was not only of interest to sociologists. The question of cutting a network into separate pieces was also of interest to computer scientists. One application is for example to create efficient parallel programs. If you execute parts of a program simultaneously, you of course want to minimise the dependency between parts that are executed concurrently. Hence, the number of links between two parts should be as small as possible. Another example is image segmentation where the network consists of similarity between neighbouring pixels, and groups in the network are formed by contiguous areas of a similar colour.

Nonetheless, finding groups in networks really took of with the work of Newman and Girvan [16]. Before that, methods of both sociologist and computer scientists alike were falling short. The sociologists' methods were not very efficient, and many methods could only be applied on a small scale, whereas the size of available data started to increase faster than ever before. The methods of the computer scientists were more efficient, but didn't seem to provide very intuitive groupings. Of course, this makes sense. Computer scientists weren't used at looking at social or biological networks, they looked at technical networks. They did not look for "natural" clusters, but just for clusters to run a program as efficiently as possible. It were these two problems that were addressed by Newman and Girvan [16].

Sociologists posed the question perhaps too broadly. They looked for all types of patterns in networks, which they termed *blockmodels* [5]. The "group" pattern, where most people know each other within a group, but not that many people outside, is only one of a whole series of possible patterns. Other patterns include for example a core-periphery structure, where core people connect amongst each other, but peripheral people only connect to the core. Another possibility is a bipartite structure, where most of the links are actually between the two groups, instead of within. All of these patterns are of course interesting in their own right, but it renders the question opaque: what exactly are you looking for in the network?

Yet the computer scientists' approach was too simplistic. You often had to specify the number of groups you wanted to find, and it assumed all groups had to be of equal size. This makes sense if you are looking to partition a network for performing parallel tasks: you know how many processors you have, and all of them should get about an equal amount of work. From the perspective of social networks, this doesn't make any sense though. We often don't know exactly how many groups to expect in a network, nor do we assume they are equally sized. In fact, it is one of the interesting questions in social networks: does the network split in two opposing factions, is there a myriad of small groups or is there no group structure at all?

The great improvement of the method of [16], which they termed *modularity*, was that you didn't have to specify the number of groups. You could simply run their community detection method, and the method would tell you how many communities there were. Of course, the more interesting patterns besides a simple group structure could not be detected, but the very focused, specific question allowed numerous researchers to work on it. Indeed, over the years, many methods were invented and tested, and we will discuss them in Chap. 2.

In general, the ingenious idea was to compare the number of links inside a group to the expected number of links. By looking for communities that maximise this difference, we could find parts of the network that are particularly well connected amongst each other. At the same time, these densely connected parts were relatively secluded from the rest of the network. This is exactly what was intuitively considered a group, or a community: it should be relatively well connected internally, and relatively well separated from the rest of the network.

It turned out that, even though it seemed to work very well, it suffered from some drawbacks. As said, one of the convenient features of modularity is that it automatically tells you how many groups there are in the network. But it turned out that modularity has a preference for relatively large groups, especially in large networks. Small groups in large networks would thus go by unnoticed. This problem is called the resolution limit, and we will address it in Chap. 3. Surprisingly, only few methods do not suffer from this problem. Only methods that are "local" in a certain sense can avoid it. But these methods cannot automatically tell you the "right" number of clusters, suggesting it is impossible to do so without a resolution limit.

Another problem of modularity is that it was thought to be an indication of group structure in networks. The value of modularity is normalised to fall between -1 and 1. It was suggested that values of 0.30 or higher would indicate a significant group structure. But such a high value of modularity could also be achieved in random graphs, casting some doubt on whether modularity could be used to say something about a significant group structure. We address this issue in Chap. 4, and suggest a solution.

To illustrate the ideas put forward, we briefly examine two applications of community detection in Chap. 6. The first focuses on finding trading communities in the international trade network of import and export. It it a long standing thesis in political science that trade reduces conflict. We show that being in the same trading community reduces conflict even more, presumably because of the high interdependency between mutual trading partners in the same trading community. Secondly, we investigate a debate network, where authors write opinion articles on the integration of minorities, and refer to each other in a positive or negative way. We show that by taking into account the valence of such references (i.e. whether they are positive or negative), community structure radically changes. By considering all references to be equal, we uncover what seem to be thematic communities: people gather around a common (sub)topic. By distinguishing negative links the more pronounced group structure becomes visible: two mutually antagonistic factions. This then brings us to the second part of the thesis. We briefly saw that Heider [11] suggested something along the lines of the ancient adage "the enemy of my enemy is my friend". Working out his ideas, Harary [4] realised that if this would hold for the entire network, it would split in two antagonistic factions. So, if everybody would play according to the ancient adage, most networks with negative links should have a relatively simple structure: they simply split in two groups. This theory became know as *social balance*, because there would be no reason for anyone to reconsider their relations.

But how would such a situation exactly come about? Suppose we start from a situation in which there is no social balance yet, then how do we get there? Perhaps somebody should change its allegiance and befriend a former enemy. But switching of position of one person might have repercussions for the rest of the network. Perhaps they too should reconsider then their allegiances. If everybody keeps doing that, will we end up in a socially balanced network? We review some models of how people change allegiances in Chap. 7, and show that some models will indeed (almost) always lead to social balance, whereas others do not.

Interestingly, this has also connections to problems of cooperation. This is a long standing problem in sociology and biology alike. In sociology, the main question is: why should somebody cooperate with me, if he can get away with cheating? In biology, the idea is similar. If a species is too "kind" to other species, he will lose the evolutionary struggle. So why should some animal cooperate with another, if he can get away with cheating? At the same time, we see cooperation all around us, at all biological scales, ranging from cooperating bacteria and cells to human societal cooperation. So how to reconcile the two?

From an evolutionary perspective, one of the most prominent explanations was put forward by Hamilton [10] and is based on kinship. Simply put: you help your sibling because the two of you share half your genes. By helping him you increase the chances of his genes surviving, and from an evolutionary perspective, this is all that matters (to some extent). Of course, cooperation is then very much based on how many genes you would share with somebody else. For single cells this is then quite a good basis for cooperation as they share most of their genes with their fellow cells. For other animals (and humans), this is restricted to nearby kin: with a cousin you only share about $1/8$ of your genes, so how much would you tend to cooperate with him?

It was suggested by Von Neumann and Morgenstern [25] that this dilemma of cooperation could be well captured in a game. In this game, you and your opponent would have two choices: either cooperate or defect. If you both cooperate, you both get €5, and if you both defect you only get €1. But, if you defect while your opponent cooperates, you would receive €8 and your opponent gets nothing. Irrespective of your opponents choice, you could better defect: if he cooperates you would get €8 instead of €5, and if he defects as well, you would get €1 instead of nothing. But what if you play multiple rounds after each other?

This lead to another possibility explanation of cooperation. In a famous experiment Axelrod [1] invited researchers to submit computer programs for playing this game as well as possible. One very simple program won the all-round tournament: tit-for-tat. This nifty little program did nothing else then cooperate if you cooperated in the last round, and defect if you defected in the last round. And to get things started, it would cooperate in the first round. Simple reciprocity seemed to beat all other strategies and cooperation could evolve because of reciprocity.

Still, this wasn't deemed enough for human cooperation. Surely, such a reciprocity effect was frequently observed, but there are also a myriad of cooperative scenarios in which people cooperate without any reciprocity. So, how could these observations be explained? One possible mechanism suggested by Nowak and Sigmund [17] was that of indirect reciprocity: if you help somebody, help will be provided to you as well, just not by the same person. This can be studied using the same game as before, but now players would change partners each round, thus preventing reciprocity. The idea of indirect reciprocity is that you should cooperate with somebody if he cooperated also with others in previous rounds. All of these strategies and mechanisms are reviewed in Chap. 9.

This finally brings us back to social balance. Indirect reciprocity necessitates to know whether somebody is cooperative or not. How would you know this if you have never seen your partner before? Simple. You ask one of your other partners. But surely, you wouldn't trust somebody that just cheated on you, so you only take advice from friends. And then we are full circle: friends of friends are friends and you should cooperate with them, while enemies of friends are also enemies, and you should defect. These dynamics are exactly the same as we studied for getting social balance. But we already know that social balance splits a network in two groups. So, even though this mechanism might lead to cooperation, it counter-intuitively also leads to a split in two groups. This might then explain the human tendency for displaying both an astonishing willingness to cooperate within their own group and an irresistible urge to exclude people from other groups.

Finally, in an online context it is also useful to know the reputation of somebody. If you meet somebody on eBay for example, should you trust him and buy that book from him? Or if you are selling you precious jewellery, should you trust the buyer to actually pay you? And how should or could you know? Of course, people can indicate whether they have concluded a deal successfully or whether there were problems. So you could use that information to get an estimate of the reputation of people. But again: why should you trust somebodies judgement if he just cheated on you? In a way, this is a recursive question: you should only trust judgements of people that are trustworthy. We will see how we can solve this issue in Chap. 10.

References

1. Axelrod R (1984) Evolution of cooperation. Basic Books, New York (ISBN 0465021220)
2. Backstrom L, Boldi P, Rosa M, Ugander J, Vigna S (2012) Four degrees of separation. In: Proceedings of the 3rd annual ACM web science conference, ACM, pp 33–42
3. Barigozzi M, Fagiolo G, Mangioni G (2011) Identifying the community structure of the international-trade multi-network. Phys A: Stat Mech Appl 390(11):2051–2066. doi:10.1016/j.physa.2011.02.004. [arXiv]1009.1731
4. Cartwright D, Harary F (1956) Structural balance: a generalization of Heider's theory. Psychol Rev 63(5):277–293. doi:10.1037/h0046049
5. Doreian P, Batagelj V, Ferligoj A (2005) Generalized blockmodeling. Cambridge University Press, Cambridge
6. Granovetter M (1973) The strength of weak ties. Am J Sociol 78:1360–1380

7. Guimerà R, Mossa S, Turtschi A, Amaral LAN (2005) The worldwide air transportation network: Anomalous centrality, community structure, and cities' global roles. Proc Nat Acad Sci USA 102(22):7794–7799. doi:10.1073/pnas.0407994102
8. Guimerà R, Nunes Amaral LA (2005) Functional cartography of complex metabolic networks. Nature 433(7028):895–900. doi:10.1038/nature03288
9. Hagmann P, Cammoun L, Gigandet X, Meuli R, Honey CJ et al (2008) Mapping the structural core of human cerebral cortex. PLoS Biol 6(7):e159. doi:10.1371/journal.pbio.0060159
10. Hamilton W (1964) The genetical evolution of social behaviour I. J Theoret Biol 7(1):1–16. doi:10.1016/0022-5193(64)90038-4
11. Heider F (1946) Attitudes and cognitive organization. J Psychol 21(1):107–112. doi:10.1080/00223980.1946.9917275
12. Kashtan N, Alon U (2005) Spontaneous evolution of modularity and network motifs. Proc Nat Acad Sci USA 102(39):13773–13778. doi:10.1073/pnas.0503610102
13. Kleinberg J, Lawrence S (2001) Network analysis. The structure of the Web. Sci (NY) 294(5548):1849–1850. doi:10.1126/science.1067014
14. Meunier D, Lambiotte R, Fornito A, Ersche KD, Bullmore ET (2009) Hierarchical modularity in human brain functional networks. Frontiers Neuroinform 3:37. doi:10.3389/neuro.11.037.2009. [arXiv]1004.3153
15. Milgram S (1967) The small world problem. Psychol Today 2(1):60–67
16. Newman M, Girvan M (2004) Finding and evaluating community structure in networks. Phys Rev E 69(2):026113. doi:10.1103/PhysRevE.69.026113
17. Nowak MA, Sigmund K (1998) Evolution of indirect reciprocity by image scoring. Nature 393(6685):573–7. doi:10.1038/31225
18. Onnela J, Saramäki J, Hyvönen J, Szabó G, Lazer D et al (2007) Structure and tie strengths in mobile communication networks. Proc Nat Acad Sci USA 104(18):7332–7336. doi:10.1073/pnas.0610245104
19. Padgett JF, Ansell CK (1993) Robust action and the rise of the medici, 1400–1434. Am J Sociol 98(6):1259–1319
20. Porter MA, Mucha PJ, Newman MEJ, Warmbrand CM (2005) A network analysis of committees in the U.S. House of Representatives. Proc Nat Acad Sci USA 102(20):7057–7062. doi:10.1073/pnas.0500191102
21. Sampson SF (1968) A novitiate in a period of change: an experimental and case study of social relationships. Ph.D. thesis, Cornell University
22. Simmel G (1950) The sociology of georg simmel, vol 92892. Simon and Schuster
23. Stouffer DB, Bascompte J (2011) Compartmentalization increases food-web persistence. Proc Nat Acad Sci USA 108(9):3648–3652. doi:10.1073/pnas.1014353108
24. Szell M, Lambiotte R, Thurner S (2010) Multirelational organization of large-scale social networks in an online world. Proc Nat Acad Sci USA 107(31):13636–13641. doi:10.1073/pnas.1004008107. [arXiv]1003.5137
25. Von Neumann J, Morgenstern O (2007) Theory of games and economic behavior. Princeton University Press, Princeton (ISBN 0691130612)
26. Watts DJ, Strogatz SH (1998) Collective dynamics of 'small-world' networks. Nature 393(June):440–442
27. Zachary W (1977) An information flow model for conflict and fission in small groups1. J Anthropol Res 33(4):452–473
28. Zhang Y, Friend A, Traud AL, Porter MA, Fowler JH et al (2008) Community structure in Congressional cosponsorship networks. Phys A: Stat Mech Appl 387(7):1705–1712. doi:10.1016/j.physa.2007.11.004

Part I
Communities in Networks

Chapter 2
Community Detection

It is clear that communities are frequently present in networks, and often have a very natural interpretation. They allow researchers to understand better the network by reducing its complexity. Our goal here is to investigate how such communities might be uncovered. We will first briefly explain the most common method for detecting communities, known as "modularity" in this chapter. We will then derive modularity from a more general framework from which some other methods can also be derived. Some of these methods have some problems, and we will discuss and analyse them in some detail, and provide some solutions in Chap. 3. For example, it remains a challenge to see how "granular" partitions should be: is it better to partition the network in many smaller communities, or in a few large communities? We address this choosing of the correct resolution in Chap. 4. If negative weights are present in network, modularity (and some variants) do not work well, and we will analyse some possible alternatives in Chap. 5. Finally, we will discuss some applications of community detection in Chap. 6.

There are two good overviews of community detection methods and algorithms. One is provided by Fortunato [16] and another by Porter et al. [39]. For a good introduction in traditional graph theory one can refer to Diestel [12], while Newman [36] provides a "complex networks" perspective. A traditional introduction into social network analysis from a sociological perspective is provided by Wasserman and Faust [50].

2.1 Modularity

Although clustering and graph partitioning have already quite a long history, they are usually not applied to (social) networks. Sociologists have constructed methods known as block modelling [13, 50], which are closer to "role[1]" detection [42] than to

[1] A role describes nodes that have similar connections to other roles, something closely related to the concept of "regular equivalence" [42, 50].

V. Traag, *Algorithms and Dynamical Models for Communities and Reputation in Social Networks*, Springer Theses, DOI: 10.1007/978-3-319-06391-1_2,

community detection. Computer scientists have been interested in graph partitioning for quite some time as well [36]. But the detection of groups in social networks really started to take off with a seminal paper by Girvan and Newman [18] in 2002. Especially their follow-up paper [37] which introduced a measure known as modularity attracted an enormous interest by a large group of researchers.

Originally, they implemented an algorithm based on the removal of edges which are part of many shortest paths [18]. The idea was that links that fall between communities are part of many such paths, because there are only few links that connect vertices from one community to another. Removing them should then disconnect the network at some point, in which case the communities should become visible. However, it was not clear at which point to stop removing edges. In order to determine this point, they introduced modularity [37]. This function should give some idea about the quality of a certain partition, and hence a clue as to when the algorithm should stop removing edges.

The idea is that communities should have relatively many edges within communities, and only little in between. Let A be an adjacency matrix of some undirected graph, so that $A_{ij} = A_{ji} = 1$ if there is an edge (i, j) and zero otherwise. Let us assume we have some fixed partition, and denote by e_{cd} the number of edges between communities c and d, corresponding to a tabulation as follows

To community

From community		1	2	$\cdots$	q	Σ
	1	e_{11}	e_{12}	$\cdots$	e_{1q}	K_1
	2	e_{21}	e_{22}	$\cdots$	e_{2q}	K_2
	$\vdots$			$\ddots$		$\vdots$
	q	e_{q1}	e_{q2}	$\cdots$	e_{qq}	K_q
	Σ	K_1	K_2	$\cdots$	K_q	$2m$

(2.1)

Then $\sum_{cd} e_{cd} = 2m$ equals twice the number of edges, since we are dealing with an undirected graph, and we count each edge twice in this manner. We are interested in $\sum_c e_{cc}/2m$ the fraction of edges within communities. Looking at this quantity, one already gets an idea of how good the partition is. However, it should be compared to how many edges we would expect to fall between two communities. This is usually done by simply taking marginals—row/column totals—which are $K_c := \sum_d e_{cd} = \sum_d e_{dc}$, the total number of edges linked to community c, as indicated in Eq. 2.1. Of course then also $\sum_c K_c = \sum_{cd} e_{cd} = 2m$. We thus arrive at the expected number of edges of $K_c K_d$ between communities c and d, which proportional to $2m$ then becomes $K_c K_d/(2m)^2$. Since we are only interested in having as many links as possible within a community we arrive at the function

$$\mathcal{Q} = \sum_c \left[\frac{e_{cc}}{2m} - \left(\frac{K_c}{2m} \right)^2 \right]. \tag{2.2}$$

The derivation provided here is quick and dirty, and we will see how a more rigorous derivation will also lead to modularity in the next section.

This measure seemed to do what was intended. Indeed when there are relatively many edges within a community, this quantity is relatively high, and approaches 1 for the most modular network possible. If a partition of a network is no better than random then $\mathcal{Q} \approx 0$. It was thought (incorrectly) that values above about 0.30 would be a sign of modular structure [37].

Although their original algorithm worked reasonably well, it was quite slow, and quickly faster algorithms appeared [8, 14, 35]. But their measure of modularity turned out to be an interesting one. Instead of using it simply to measure how well the network was partitioned, people began to optimize the measure itself [14, 21, 38]. However, it has some deficits and problems, which we will discuss in the next chapter. But first we will derive this measure of modularity in a more general framework, and go over some of the other possible methods for community detection.

2.2 Canonical Community Detection

In this chapter we will derive modularity in a more general setting, starting from first principles, similar to Reichardt and Bornholdt [41]. As stated, this more general framework will be used throughout the thesis, and forms the backbone of our analysis. Although not all methods can be represented in this way, it is a reasonably general framework, and we therefore refer to it as the canonical community detection framework.

Let us first start with some basic notation. Let $G = (V, E)$ be an undirected graph with nodes $V = \{1, \ldots, n\}$ and $E = \{(i, j): i, j \in V\}$ the undirected edges of the graph G. Furthermore, we denote by A the adjacency matrix of G, such that $A_{ij} = 1$ if there is an (i, j) link, and $A_{ij} = 0$ otherwise. For an undirected graph the adjacency matrix $A = A^\top$ is symmetric where $A^\top$ denotes the transpose (i.e. $A^\top_{ji} = A_{ij}$). In addition, each link might have an associated weight $w_{ij} \in \mathbb{R}$, which we assume to be positive for the moment (we will consider the possibility of negative weights explicitly in Chap. 5). It might sometimes be useful to have a weighted adjacency matrix where $A_{ij} = w_{ij}$ when there is an (i, j) link. If we use the weighted adjacency matrix, this will be stated explicitly. The unweighted case then also corresponds to a weight of $w_{ij} = 1$. We denote the partition by $\sigma_i \in \{1, \ldots, q\}$ where each σ_i indicates the community to which node i belongs, so σ is the membership vector. Alternatively, it is sometimes useful to denote communities as sets of nodes. We will use $\mathcal{C} = \{C_1, C_2, \ldots, C_q\}$ to denote the set of community sets, such that each set $C_c = \{i \in V \mid \sigma_i = c\}$ contains the nodes which belong to community c. Any partition of the graph is assumed to be non-overlapping and complete. Stated differently, every node belongs to a single community, in other words, for any valid partition it holds that $\bigcup_{c=1}^q C_c = V$ (all nodes are in a community) and $C_c \cap C_d = \emptyset$ for $c \neq d$ (no node is in more than one community). The size of a community (the number of nodes in a community) will usually be denoted by $n_c = |C_c|$. When referring to "the partition" this might be either to σ or to $\mathcal{C}$, and should be clear from context. We will mostly focus on undirected and unweighted graphs, but most of these quantities can be straightforwardly extended to directed and weighted graphs.

Although the overall objective—detect communities—might be clear, what exactly constitutes a community is not undisputed. For example, one can take into account the number of triangles within a community, the size of the largest clique, or k-connectedness, and so forth. For example, traditional clustering works with notions of distance $d(i, j)$ between node i and j [51]. We shall start from a first principle basis that is due to Reichardt and Bornholdt [41]. The basic idea is to only specify the general framework, which can be made more specific, for example by counting the number of triangles or common neighbours. A commonly accepted idea of a community is that it should be a relatively dense subgraph that is relatively well separated from the rest of the graph. This means there should be relatively:

1. many present links within communities;
2. few absent links within communities;
3. few present links between communities; and
4. many absent links between communities.

Taking these assumptions, we reward present links (a_{ij}) and punish absent links (b_{ij}) within communities, while we punish present links (c_{ij}) and reward absent links (d_{ij}) between communities. Summarizing, we have the following weights:

	$A_{ij} = 1$	$A_{ij} = 0$
$\delta(\sigma_i, \sigma_j) = 1$	a_{ij}	$-b_{ij}$
$\delta(\sigma_i, \sigma_j) = 0$	$-c_{ij}$	d_{ij}

where all weights $a_{ij}, b_{ij}, c_{ij}, d_{ij} \geq 0$ remain to be specified and δ is the Kronecker delta

$$\delta(a, b) = \begin{cases} 1 & \text{if } a = b \\ 0 & \text{if } a \neq b \end{cases} \tag{2.3}$$

so that $\delta(\sigma_i, \sigma_j) = 1$ if $\sigma_i = \sigma_j$ both i and j are in the same community, and 0 otherwise. We then denote by $\mathcal{H}$ the objective function

$$\begin{aligned}\mathcal{H}(\sigma) = -\sum_{ij} & \Big[a_{ij} A_{ij} - b_{ij}(1 - A_{ij})\Big]\delta(\sigma_i, \sigma_j) \\ & + \Big[-c_{ij} A_{ij} + d_{ij}(1 - A_{ij})\Big](1 - \delta(\sigma_i, \sigma_j)).\end{aligned}$$

The minus sign is only a matter of convention, and in this case we would like to minimize this function. The optimization problem is then

$$\min_{\sigma} \mathcal{H}(\sigma), \tag{2.4a}$$

$$\text{s.t. } \sigma \in \{1, \ldots, q\}^n. \tag{2.4b}$$

We will refer to $\mathcal{H}(\sigma)$ as the cost of a partition σ, and so the optimal partition has minimal cost. Now if we suppose that links within communities should be equally rewarded/punished as links between communities, i.e. $a_{ij} = c_{ij}$ and $b_{ij} = d_{ij}$, we can simplify to

$$\mathcal{H}(\sigma) = -\sum_{ij} a_{ij} A_{ij}(2\delta(\sigma_i, \sigma_j) - 1) - b_{ij}(1 - A_{ij})(2\delta(\sigma_i, \sigma_j) - 1).$$

Since we are looking for the minimum of $\mathcal{H}(\sigma)$ we can remove factors that do not depend on σ, i.e. not depending on $\delta(\sigma_i, \sigma_j)$. Furthermore, any multiplication with a constant leaves the minimum unchanged. Using these observations, we can simplify to

$$\mathcal{H}(\sigma) = -\sum_{ij}(a_{ij} A_{ij} - b_{ij}(1 - A_{ij}))\delta(\sigma_i, \sigma_j). \tag{2.5}$$

This is the objective function we will analyse in this thesis, and forms the core of our enquiry. The weights a_{ij} and b_{ij} remain to be specified, but are assumed to be non-negative $a_{ij}, b_{ij} \geq 0$.

Irrespective of the specific weights chosen, any community should be connected. To show this, assume on the contrary there is a community C which is disconnected, so that for some partition $C = S \cup S'$, with $S \cap S' = \emptyset$, there are no edges from S to S'. In that case, if we split the community into S and S', we decrease the cost function assuming there is at least one $b_{ij} > 0$, so that it cannot be optimal.

Different choices for the weights a_{ij} and b_{ij} lead to different methods for community detection. For example, we could imagine taking into account the number of common neighbours between i and j for absent links, so that $b_{ij} = |N(i) \cap N(j)|$, or the number of independent paths between i and j, similar to the original algorithm of Girvan and Newman [18]. Numerous choices could be made, and we will review some of the possibilities (for an overview, refer to Table 2.1).

2.2.1 Reichardt and Bornholdt

One choice consists of comparing the original network to a randomized network, a random null model, as considered by Reichardt and Bornholdt [41]. Let us assume the probability for a link is p_{ij}, which we will specify later. The weight of a missing link is $b_{ij} = \gamma_{\mathrm{RB}} p_{ij}$, while the weight of a present link is $a_{ij} = w_{ij} - b_{ij}$, where w_{ij} is the weight of the (i, j) link, or $w_{ij} = 1$ if the graph is not weighed and γ_{RB} a parameter used to weigh the importance of the randomized network. Summarizing, the weights are

$$a_{ij} = w_{ij} - \gamma_{\text{RB}} p_{ij}, \tag{2.6a}$$

$$b_{ij} = \gamma_{\text{RB}} p_{ij}. \tag{2.6b}$$

In other words, whenever a link has more weight than expected in the randomized network, we reward that link if it is within a community. Including a missing link in a community would be punished slightly if the expected weight of a link is low. Working out this choice leads to

$$\begin{aligned}\mathcal{H}_{\text{RB}} &= -\sum_{ij} \left[(w_{ij} - \gamma_{\text{RB}} p_{ij}) A_{ij} - \gamma_{\text{RB}} p_{ij} (1 - A_{ij}) \right] \delta(\sigma_i, \sigma_j) \\ &= -\sum_{ij} \left[w_{ij} A_{ij} - \gamma_{\text{RB}} p_{ij} \right] \delta(\sigma_i, \sigma_j) \end{aligned} \tag{2.7}$$

In the following we will assume that the graph is unweighted and that $w_{ij} = 1$. We can rewrite Eq. (2.7) slightly to gain some additional insight. We gather the terms per community, and arrive at

$$\begin{aligned}\mathcal{H}_{rb} &= -\sum_{ij} (A_{ij} - \gamma_{\text{RB}} p_{ij}) \delta(\sigma_i, \sigma_j) \\ &= -\sum_{c} \sum_{ij} (A_{ij} - \gamma_{\text{RB}} p_{ij}) \delta(\sigma_i, c) \delta(\sigma_j, c).\end{aligned}$$

So if we write

$$e_c = \sum_{ij} A_{ij} \delta(\sigma_i, c) \delta(\sigma_j, c)$$

for the number[2] of edges in community c and

$$\langle e_c \rangle_{p_{ij}} = \sum_{ij} p_{ij} \delta(\sigma_i, c) \delta(\sigma_j, c)$$

for the expected number of edges in community c, we can rewrite Eq. (2.7) as

$$\mathcal{H}_{rb} = -\sum_{c} \left[e_c - \gamma_{\text{RB}} \langle e_c \rangle_{p_{ij}} \right].$$

In general, the average of some quantity will usually be denoted by $\langle \cdot \rangle$. In other words, this objective function considers the difference between the actual number of edges within a community and the expected number of edges within a community given a random null model. Hence, there are two ways for improving this function: by having more edges within a community, or by having less expected edges within

[2] Technically twice the number of edges in community c for undirected graphs.

a community. The expected edges weigh more heavily with higher γ_{RB}, so that it effectively constrains the community sizes. But we will get back to this later on.

Various random null models can be chosen to specify p_{ij}. One possibility is to take a simple Erdös-Renyí (ER) graph [5] where each link[3] appears with the same probability $p = m/n^2$, where $m = |E|$ the number of edges and n the number of nodes. We then set

$$p_{ij} = p = \frac{m}{n^2}.$$

The expected number of edges within a community is then simply

$$\langle e_c \rangle_p = pn_c^2$$

where n_c is the number of nodes of community c. In this case the density within a community is expected to be about the same as the density of the graph in general. The objective function as a sum over communities then simplifies to

$$\mathcal{H}_{\text{RB}} = \sum_c \left[e_c - \gamma_{\text{RB}} pn_c^2 \right].$$

However, an ER graph is not realistic in the sense that the degree $k_i = \sum_j A_{ij}$ of a node deviates from what is empirically expected. An ER graph has a Poissonian degree distribution so that

$$\Pr(k) = \frac{\langle k \rangle^k e^{-\langle k \rangle}}{k!},$$

while in reality the degree distribution is highly skewed and heavy tailed, and follows more a power law [36]

$$\Pr(k) \sim k^{-\tau}.$$

So, another common null model is the configuration model, which takes into account the degree. A simple way to construct a randomized network with the same degrees is to cut all links in half, so that each link has k_i stubs (one half of a link), and to connect all the stubs randomly. We then arrive at the expected number of links between i and j of

$$p_{ij} = \frac{k_i k_j}{2m}. \tag{2.8}$$

The derivation of the quantity is as follows. We have k_i ways to choose a stub from node i, since it has k_i stubs to connect. Similarly, we have k_j ways for choosing to connect to node j. Finally, we choose from $2m$ stubs (twice for each link). The expected number of links within a community is then

[3] We here include the possibility of self-loops.

$$\langle e_c \rangle_{\text{conf}} = \frac{K_c^2}{2m}, \tag{2.9}$$

where $K_c := \sum_i k_i \delta(\sigma_i, c)$ is the sum of the degrees of the nodes in community c. If the total degree is relatively high, we expect more edges to fall within the community. Notice that this no longer corresponds to the density of a community. The objective function becomes

$$\mathcal{H}_{\text{RB}} = \sum_c \left[e_c - \gamma_{\text{RB}} \frac{K_c^2}{2m} \right]. \tag{2.10}$$

The classical modularity can then be derived by taking $\gamma_{\text{RB}} = 1$, using the configuration model, and normalize by $\frac{1}{2m}$ and inverse the sign to arrive at

$$\mathcal{Q} = \frac{1}{2m} \sum_{ij} \left(A_{ij} - \frac{k_i k_j}{2m} \right) \delta(\sigma_i, \sigma_j). \tag{2.11}$$

or as a sum over communities, which is sometimes easier to use,

$$\mathcal{Q} = \sum_c \left[\frac{e_c}{2m} - \left(\frac{K_c}{2m} \right)^2 \right], \tag{2.12}$$

and we retrieve the definition provided in Eq. (2.2).

2.2.2 *Arenas, Fernández and Gómez*

A particular problem of modularity (and the RB model in general) is the so-called resolution limit, which we will analyse more in-depth later on (see Chap. 3). The basic problem in the resolution limit is that communities are merged together while they actually shouldn't. This problem can be addressed to a certain extent by the resolution parameter γ_{RB} in the RB model, but other solutions have been proposed. One noteworthy solution by Arenas et al. [2] (AFG) consists of adding self-loops to nodes so as to prevent these nodes from being merged. In other words, they use almost the same weights as RB, but then adapted for the added self-loops of strength γ_{AFG}. This idea translates into the weights

$$a_{ij} = w_{ij} - b_{ij}, \tag{2.13a}$$

$$b_{ij} = p_{ij}(\gamma_{\text{AFG}}) - \gamma_{\text{AFG}} \delta_{ij}, \tag{2.13b}$$

where $\delta_{ij} = \delta(i, j) = 1$ if $i = j$ and zero otherwise. The authors use the classical configuration model for the null-model, and use

$$p_{ij}(\gamma_{\mathrm{AFG}}) = \frac{(k_i + \gamma_{\mathrm{AFG}})(k_i + \gamma_{\mathrm{AFG}})}{2m + n\gamma_{\mathrm{AFG}}}. \tag{2.14}$$

Their model then becomes (up to multiplicative scaling)

$$\mathcal{H}_{\mathrm{AFG}}(\sigma) = -\sum_{ij} \left(A_{ij}w_{ij} + \gamma_{\mathrm{AFG}}\delta_{ij} - p_{ij}(\gamma_{\mathrm{AFG}})\right)\delta(\sigma_i, \sigma_j) \tag{2.15}$$

which is simply Eq. (2.7) with self-loops added. The benefit of this method is that it leaves unchanged properties that depend on the eigenvectors or on the difference of the eigenvalues. In order to see that, observe that we could also have transformed the original matrix A to $A' = A + \gamma_{\mathrm{AFG}} I_n$ where I_n is the $n \times n$ identity matrix, i.e. $I_n = \mathrm{diag}(1, \ldots, 1)$. Now suppose that λ is an eigenvalue and v the corresponding eigenvector of A (i.e. $Av = \lambda v$), then also $A'v = Av + \gamma_{\mathrm{AFG}} I_n v = (\lambda + \gamma_{\mathrm{AFG}})v$ so that v is an eigenvector of A' and $\lambda + \gamma_{\mathrm{AFG}}$ an eigenvalue of A'. Although the same idea could be investigated using the ER null model this has not been considered. Notice that the AFG model is indeed different from the RB null-model and that the two are only equal for $\gamma_{\mathrm{AFG}} = 0$ and $\gamma_{\mathrm{RB}} = 1$ in general.

2.2.3 Ronhovde and Nussinov

Ronhovde and Nussinov [43] (RN) do not include any null model, in order to avoid issues with the resolution limit, and in general set

$$a_{ij} = w_{ij}, \tag{2.16a}$$

$$b_{ij} = \gamma_{\mathrm{RN}}, \tag{2.16b}$$

(although for specific networks, such as with negative weights, they allow some minor changes). Working this out we obtain

$$\mathcal{H}_{\mathrm{RN}}(\sigma) = -\sum_{ij} (A_{ij}(w_{ij} + \gamma_{\mathrm{RN}}) - \gamma_{\mathrm{RN}})\delta(\sigma_i, \sigma_j). \tag{2.17}$$

Notice that for unweighted graphs (i.e. $w_{ij} = 1$) up to rescaling this is equal to

$$\mathcal{H}_{\mathrm{RN}}(\sigma) = -\sum_{ij} \left(A_{ij} - \frac{\gamma_{\mathrm{RN}}}{1 + \gamma_{\mathrm{RN}}}\right)\delta(\sigma_i, \sigma_j). \tag{2.18}$$

If we compare this to the RB model with an ER null model, the RN model is equal to the RB model if

$$\gamma_{\mathrm{RN}} = \frac{1 - \gamma_{\mathrm{RB}} p}{\gamma_{\mathrm{RB}}}.$$

For weighted graphs, the models are not necessarily the same however.

2.2.4 Constant Potts Model

A formulation that also has no null model, similar to Ronhovde and Nussinov [43], but which resembles more closely the RB model is provided by

$$a_{ij} = w_{ij} - b_{ij}, \tag{2.19a}$$
$$b_{ij} = \gamma_{\text{CPM}}, \tag{2.19b}$$

which results in

$$\mathcal{H}_{\text{CPM}} = -\sum_{ij}(A_{ij}w_{ij} - \gamma_{\text{CPM}})\delta(\sigma_i, \sigma_j). \tag{2.20}$$

We call this the Constant Potts Model because it only compares the network to a constant parameter γ_{CPM} [49].

As can be expected, this model is rather similar to the RN model and the RB model. The RB and RN model are equivalent if $\gamma_{\text{CPM}} = \gamma_{\text{RB}}\, p$ and the ER null model is used. The RN model is only equal to the CPM model for unweighted graphs, in which case we have $\gamma_{\text{CPM}} = \frac{\gamma_{\text{RN}}}{1+\gamma_{\text{RN}}}$.

2.2.5 Label Propagation

Finally, the label propagation (LP) method [40] can be shown to be equivalent to the Potts model [48]

$$a_{ij} = w_{ij}, \tag{2.21a}$$
$$b_{ij} = 0. \tag{2.21b}$$

which results in the trivially optimized

$$\mathcal{H}_{LP} = -\sum_{ij} A_{ij}w_{ij}\delta(\sigma_i, \sigma_j) \tag{2.22}$$

This model is equivalent to the RB model, the RN model and CPM as long as $\gamma_{\text{RB}} = \gamma_{\text{RN}} = \gamma_{\text{CPM}} = 0$. This is the least interesting formulation, since there is only one global optimum, namely all nodes belong to a single community, which is trivial. However, the local minima could be of some interest. Furthermore, these local minima can be relatively quickly found, rendering the complexity of the associated algorithm essentially linear [40].

Table 2.1 Overview of different methods

Method	a_{ij}	b_{ij}	Objective function
Modularity (p. 11)	$w_{ij} - \frac{k_i k_j}{2m}$	$\frac{k_i k_j}{2m}$	$\frac{1}{2m}\sum_{ij}\left(A_{ij}w_{ij} - \frac{k_i k_j}{2m}\right)$
RB (p. 15)	$w_{ij} - \gamma_{\text{RB}} p_{ij}$	$\gamma_{\text{RB}} p_{ij}$	$-\sum_{ij}(A_{ij}w_{ij} - \gamma_{\text{RB}} p_{ij})\delta(\sigma_i, \sigma_j)$
AFG (p. 18)	$w_{ij} - b_{ij}$	$p_{ij}(\gamma_{\text{AFG}}) - \gamma_{\text{AFG}}\delta_{ij}$	$-\sum_{ij}(A_{ij}w_{ij} + \gamma_{\text{AFG}}\delta_{ij} - p_{ij}(\gamma_{\text{AFG}}))\delta(\sigma_i, \sigma_j)$
RN (p. 19)	w_{ij}	γ_{RN}	$-\sum_{ij}(A_{ij}(w_{ij} + \gamma_{\text{RN}}) - \gamma_{\text{RN}})\delta(\sigma_i, \sigma_j)$
CPM (p. 20)	$w_{ij} - \gamma_{\text{CPM}}$	γ_{CPM}	$-\sum_{ij}(A_{ij}w_{ij} - \gamma_{\text{CPM}})\delta(\sigma_i, \sigma_j)$
LP (p. 20)	w_{ij}	0	$-\sum_{ij} A_{ij}w_{ij}\delta(\sigma_i, \sigma_j)$

2.2.6 Random Walker

There are also some other derivations of modularity (and some of the others models) in terms of a random walk on a graph, by Delvenne et al. [11]. They focus on the time it takes for a random walker to escape from a community. Since a random walker should be trapped within a community for a considerable time, if we try to maximize how long the walker will remain in the same community, we should find communities.

Let us take a look at how we can represent such a random walk on a graph. Suppose we start our walk with a certain probability $\pi(0)$ in some node, so that $\pi_i(0)$ gives the probability we start in node i. The random walker simply follows each link with uniform probability. So, from a node i, it follows the link (i, j) with probability $1/k_i$. If we define $M = (D^{-1}A)^\top$ where $D = \text{diag}(k_1, k_2, \ldots, k_n)$ has the degrees on the diagonal, then M_{ij} gives the transition probabilities for moving from node i to j. The probability we are in a certain node after a single step is then $\pi(t+1) = M\pi(t)$, and so $\pi(t) = M^t\pi(0)$. If we assume the network to be (strongly) connected and aperiodic, this matrix is primitive, and according to the Perron-Frobenius theorem, in the limit

$$\lim_t \pi(t) = \pi = M\pi \tag{2.23}$$

this probability becomes stationary, and π is the dominant eigenvector of M. So, after a sufficient long time, each node will be visited with probability π_i.

Now let us give each node some label σ_i. Suppose the random walker records the labels σ_i of nodes visited in a random variable X_t, so that if the random walker was in node i after t steps, then $X_t = \sigma_i$. As stated, we would like to know whether the random walker remains in the same community for a long time. Suppose that the label σ_i of a node indicates the community. If a random walker stays within the same community, the random variable X_t is likely to be the same. This can be measured through the autocovariance between X_τ and $X_{\tau+t}$ with $t > 0$, which is defined as

$$\mathrm{Cov}(X_\tau, X_{\tau+t}) = \mathbb{E}(X_\tau X_{\tau+t}) + \mathbb{E}(X_t)^2.$$

The expected value of X_t can be easily calculated, if we assume the random walk to become stationary. In that case, $\mathbb{E}(X_t) = \sum_i \sigma_i \pi_i = \sigma^\top \pi = \pi^\top \sigma$, and so

$$\mathbb{E}(X_t)^2 = (\sigma^\top \pi)(\pi^\top \sigma) = \sigma^\top \pi \pi^\top \sigma.$$

To calculate $\mathbb{E}(X_\tau X_{\tau+t})$ at stationarity we obtain that

$$\mathbb{E}(X_t X_{t+\tau}) = \sum_i \sigma_i \pi_i (M^t \sigma)_i = \sigma^\top \Pi M^t \sigma,$$

where $\Pi = \mathrm{diag}(\pi)$. We encode $\sigma = S\alpha$ where $\alpha = (1, \ldots, q)$ and S is the $n \times q$ community matrix, such that $S_{ic} = 1$ if $\sigma_i = c$ node i is in community c and 0 otherwise (see also Sect. 2.3.4). We can then write the covariance as

$$\mathrm{Cov}(X_\tau, X_{\tau+t}) = \alpha^\top R(t)\alpha$$

where

$$R(S, t) = S^\top (\Pi M^t - \pi^\top \pi) S$$

is the so called stability matrix. Each element $R(S, t)_{cd}$ denotes the probability to start in community c and go to community d after t steps minus the probability two random walkers are in c and d. Since we are interested in maximizing the time spent inside a community, we would like to maximize $R(S, t)_{cc}$. In other words, we would like to find $\max_S \mathrm{Tr} R(S, t)$ where $\mathrm{Tr} X = \sum_i X_{ii}$ is the trace of some matrix X. However, we should remain within the community for all time up to t. So we define the stability of a partition S at time t as

$$r(S, t) = \min_{\tau < t} \mathrm{Tr} R(S, \tau).$$

and we would like to maximize this $r(S, t)$ for some t. In general, we can write

$$\mathrm{Tr} R(S, t) = \sum_c S_c^\top (\Pi M^t - \pi^t \pi) S_c = \sum_{ij} (\Pi (M^t)_{ij} - \pi_i \pi_j) \delta(\sigma_i, \sigma_j).$$

If the random walk is undirected, we have that $\pi_i = \frac{k_i}{2m}$. Now suppose we look at only a single step, or $t = 1$, so that we obtain that

$$\mathrm{Tr} R(S, 1) = \frac{1}{2m} \sum_{ij} \left(A_{ij} - \frac{k_i k_j}{2m} \right) \delta(\sigma_i, \sigma_j).$$

Hence, we recover exactly modularity for time $t = 1$ on undirected networks. For directed networks this quantity differs from the null model originally proposed for directed networks [32]. Approximating this equation around $t = 1$, a different interpretation of the resolution parameter for the RB model is obtained, namely that $\gamma_{\mathrm{RB}} \approx 1/t$. However, this only holds approximately. Furthermore, some related type of (continuous time) random walk gives an alternative derivation for the RB model with an ER null model [28].

2.2.7 *Infomap*

A quite successful method that unfortunately doesn't fit within this framework is Infomap [44, 45]. We include a brief description of this method since it is one of the best performing methods outside of this framework, although certainly not the only one (see [1, 31]). It is based on ideas of information theory, which we will briefly explain. Information theory concerns itself with the representation of information, and naturally involves also the compression of information. For example, if we have a very long piece of text which reiterates "Help! Help! Help! Help! Help!", it would be more efficient to simply write "Help! (5×)". In a similar fashion, one can imagine being able to compress other information, which these days is often used when creating `.zip` files, but also in videos (`.mp4`), images (`.jpg`) or music (`.mp3`).

Infomap focuses on trying to compress the list of nodes visited by a random walker on a graph. We record all the nodes a walker has visited, for example "1, 5, 3, 2", meaning that the walker first visited node 1 then 5, then 3 and finally 2, similar to the random variable X_t in the previous section. If we continue this walk for a very long time, we expect him to spend a reasonable amount of time in the same community. We may use this to represent the list of all nodes the walker has visited in a more efficient way. Hence, the idea of a random walker is similar to the previous section, although the objective is different: previously the focus was on staying in the same community as long as possible, while here the focus is on having a description of the random walk which is as short as possible.

Let us first briefly review the basics of information theory.

Information Theory

Information theory mostly deals with how information can be represented and quantified [9, 33]. The information value of a certain event is logarithmically inverse to the probability of it occurring. In other words, suppose that X is a random variable and that $\Pr(X = x) = p(x)$, then the information associated with event x is

$$I(x) = \log \frac{1}{p(x)} = -\log p(x). \tag{2.24}$$

This has two nice properties: (1) the information associated with two independent identically distributed events x is then $2I(x)$ so contains twice the information; and (2) if x is sure to happen, so when $p(x) = 1$, it contains no information and $I(x) = 0$. The maximum information about a certain event is then when $p(x) \to 0$, which makes sense. After all, if x happens almost never, it provides much information when it actually does happen.

Given a certain distribution $p(x)$ we can also ask what is the expected information associated with the random variable X. This measure is also known as the entropy, and can be written as

$$H(X) = \mathbb{E}(I(X)) = -\sum_x p(x) \log p(x). \tag{2.25}$$

If we look at the probability of X given Y, or $\Pr(X = x \mid Y = y) = p(x \mid y)$, the information content associated to x given y is then $I(x \mid y) = -\log p(x \mid y)$. The entropy of $H(X \mid Y = y)$ is then

$$H(X \mid Y = y) = -\sum_x p(x \mid y) \log p(x \mid y),$$

hence the conditional entropy is

$$\begin{aligned} H(X \mid Y) = \mathbb{E}(H(X \mid Y = y)) &= -\sum_y p(y) \sum_x p(x \mid y) \log p(x \mid y) \\ &= -\sum_{xy} p(x, y) \log \frac{p(x, y)}{p(y)}. \end{aligned} \tag{2.26}$$

Notice that if Y and X are independent random variables, then $H(X \mid Y) = H(X)$, and otherwise $H(X \mid Y) \leq H(X)$. In other words, conditioning always decreases the entropy. Furthermore, if X is completely determined by Y then $H(X \mid Y) = 0$, which makes sense since knowing Y we also know X. Similarly, the joint entropy can be defined as

$$H(X, Y) = -\sum_{xy} p(x, y) \log p(x, y), \tag{2.27}$$

and hence

$$\begin{aligned} H(X, Y) = H(Y, X) &= H(Y \mid X) + H(X), \\ &= H(X \mid Y) + H(Y). \end{aligned}$$

If X and Y are independent random variables then $H(X \mid Y) = H(X)$, and so $H(X, Y) = H(X) + H(Y)$. Since $H(X \mid Y) \geq 0$, we have $H(X, Y) \geq H(X)$ and $H(X, Y) \geq H(Y)$, and so the joint entropy is always larger than the entropy of a single random variable.

Now suppose we wish to represent a series of random variables, which are independently identically distributed (iid) with distribution $p(x)$. In this context it is common to talk about symbols and a code to represent that symbol. For example, suppose that our distribution gives the symbol a with probability p_a and b with probability p_b, and likewise p_c and p_d. We will usually represent codes of symbols in binary code, and so we can represent the symbols by using the following code.

Symbol	Code
a	00
b	01
c	10
d	11

Here the code length $b_i = 2$ for all codes i. So, the code for the sequence "$adba$" is then "00110100". However, if we know that some symbols occur more often then others, we might want to assign shorter codes to symbols that are more often used. For example if the symbols occur with probabilities $p_a = 0.6$, $p_b = 0.2$ and $p_c = p_d = 0.1$, we could use the following codes.

Symbol	Code
a	0
b	10
c	110
d	111

Notice that the code for a is shorter $b_a = 1$, but the codes for c and d are longer, $b_c = b_d = 3$. The code for the same sequence as before is now "0111100", which has a total length of 7 bits, while the original code used 8 bits. Notice that we can identify the codes unambiguously, because no code appears in the beginning of another code, a property known as prefix-free. In general, if we look at the expected code length per symbol, this is

$$\sum_i p_i b_i = 0.6 \cdot 1 + 0.2 \cdot 2 + 0.1 \cdot 3 + 0.1 \cdot 3 = 1.6$$

using the adapted code, while for the original codes this was $\sum_i p_i b_i = 2$. So, we improved the representation of this sequence by changing the codes. The idea is now that the number of possibilities for a codeword of a length b_i should be inversely proportional to its probability, so that $2^{b_i} = 1/p_i$, or the number of bits[4] $b_i = -\log p_i$. Rare symbols then get long codes, and often occurring symbols shorter codes. The expected code length per symbol is then

[4] This could be expressed in a different base as well. Since the base only changes the properties up to a multiplicative constant, we ignore this and simply take the natural logarithm.

$$\sum_i p_i b_i = -\sum_i p_i \log p_i = H(X).$$

The amazing thing is that this is also the optimal code length per symbol. In other words, we cannot represent the information in a shorter code per symbol than the entropy. This is known as the famous Shannon source-coding theorem [9]. The actual codes attaining this bound are known as Huffman codes. For our purposes here, we do not need this machinery, and we will not discuss it further.

Compressing Random Walks

How can we use compression to find communities? As stated, we expected a random walker to remain in the same community for a substantial amount of time. The ingenious idea is then that as long as we remain in the same community we can use shorter codes for nodes in the same community. That is, we can use the same code for two different nodes in two different communities. Compare it to calling somebody on a land line. If you need to call someone within the same village (or even organisation) you usually only need a few numbers. For example, you dial your best friend with the phone number "1105". Now if you want to call somebody in another village (with number "38"), you will first have to dial out (using the code "0"), then dial the access code and then the phone number again. For example, your other friend lives in another town and you dial "0-38-1105". Notice that the actual phone number can be the same for both friends: "1105". This is the same idea for the random walker: nodes in different communities can reuse the same code.

If we do not consider any partition, by Shannon's source coding theorem, we can represent the list of nodes visited with $H(X) = -\sum_i \pi_i \log \pi_i$ bits per step, where π_i are the stationary probabilities of the random walker as derived in Eq. (2.23). If we do consider a partition σ, we can reuse the same codes for nodes in different communities, which should shorten the average code length for that community.

The probability that a random walker stays within a community is then

$$\rho_c = \frac{e_c}{K_c}$$

where $e_c = \sum_{ij} A_{ij}\delta(\sigma_i, \sigma_j)$ the total number of edges as before, and $K_c = \sum_i k_i \delta(\sigma_i, c)$ the total degree. The probability to leave a community is then of course $1 - \rho_c$. The probability a certain community is visited is then

$$q_c = \sum_i \pi_i \delta(\sigma_i, c).$$

We should also define a code for moving outside a community to another community, similar to dialling a "0" for dialling out. We include this code for exiting from community c in the entropy, in order to take it into account. The entropy for moving within a community c (or exiting) is then

$$H_c = -\sum_i \frac{\pi_i}{q_c + (1-\rho_c)} \log \frac{\pi_i}{q_c + (1-\rho_c)}$$
$$- \frac{1-\rho_c}{q_c + (1-\rho_c)} \log \frac{1-\rho_c}{q_c + (1-\rho_c)},$$

so that we can choose optimal codes of average code length H_c for that community. In addition, if the random walker exits from a community, the average code length for indicating to which community the random walker goes is then

$$H_q = -\sum_c q_c \log q_c$$

With probability q_c we then incur the average code length of H_c while with probability $(1-\rho) := \sum_c (1-\rho_c)$ we incur the cost of switching communities. So, the total expected code length is then

$$\mathrm{L}(\sigma) = (1-\rho)H_q + \sum_c q_c H_c. \tag{2.28}$$

This is known as the map equation, and we try to minimize this expected code length. The derivation here is slightly different from the original [44], but is similar in spirit. Unlike the other models, we will not analyse this model in great detail, but it is included for the sake of completeness.

2.2.8 Alternative Clustering Methods

As stated earlier, the approach of community detection is somewhat recent, and different approaches have been used before. There exists a multitude of general clustering techniques, such as hierarchical clustering or k-means clustering, which are usually applied to datasets in some Euclidean space [15, 23, 27, 51]. By using some graph similarity (or distance) type of measure, it is possible to apply these existing techniques on graphs [46]. Hierarchical clustering for example merges two groups depending on the similarity of the two groups (taking a greedy outlook), thus resulting in a dendrogram of merges. The k-means method tries to iteratively minimize the average within cluster distances by minimizing the distance to some cluster average.

Similarities between nodes can be derived in many different ways. One such similarity measure can for example be derived by considering the expected commuting time to go from node i to node j in a random walk on a graph [52]. This can be based on the graph Laplacian, which is defined as

$$\mathcal{L} = D - A, \tag{2.29}$$

where A is the adjacency matrix and $D = \text{diag}(k_1, \ldots, k_n)$ is the diagonal degree matrix. Notice that

$$\begin{aligned} u^\top \mathcal{L} u &= \sum_{ij} u_i L_{ij} u_j \\ &= \sum_{ij} [u_i \delta_{ij} k_i u_j] - \sum_{ij} [u_i A_{ij} u_j] \\ &= \sum_{ij} A_{ij} (u_i - u_j)^2 \end{aligned}$$

so that $\mathcal{L}$ is positive-semidefinite and has only non-negative eigenvalues. We won't go into the details, but the expected commuting time C_{ij} to go from node i to node j can be expressed as [17]

$$C_{ij} = 2m(e_i - e_j)\mathcal{L}^+(e_i - e_j) \tag{2.30}$$

where e_i is the ith basis vector and L^+ is the pseudo inverse of the Laplacian

$$\mathcal{L}^+ = \left(\mathcal{L} - \frac{1}{n}\right)^{-1} + \frac{1}{n}. \tag{2.31}$$

It can be proven that C_{ij} is a proper distance metric, which can then be used in other clustering techniques for further processing.

Another approach also based on the Laplacian is that of spectral graph partitioning (for details, see [3, 36]). This idea is based on trying to minimize the cut-size. Assume we have some vector $s \in \{-1, 1\}^n$, where $s_i = -1$ indicates node i is in group 1 and if $s_i = 1$ it is in group 2. Then the total number of edges running between the two groups can be written as

$$\sum_{ij} A_{ij} \frac{1}{2}(1 - s_i s_j) = \frac{1}{2} s^\top \mathcal{L} s. \tag{2.32}$$

Realising that $\frac{1}{2}(1 - s_i s_j) = 1 - \delta(\sigma_i, \sigma_j)$, we then recognize the trivially optimized label propagation method (LP) from Eq. 2.22. The trivial solution is simply $s = (1, \ldots, 1)$ in which case $s^\top \mathcal{L} s = 0$. That is why often in this context an additional constraint is imposed, namely that the two groups should be of roughly equal size. Solving this leads to the eigenvector u_2 corresponding to the second-smallest eigenvalue λ_2 of the Laplacian $\mathcal{L}$, and setting $s_i = \text{sgn}(u_{2i})$. This eigenvector is also known as the Fiedler vector. The first eigenvalue $\lambda_1 = 0$, and the second eigenvalue λ_2 is only zero if the graph is disconnected. For this reason, it is also known as the algebraic connectivity. There are also other variants of spectral graph partitioning, for example based on the normalized Laplacian $D^{-1}\mathcal{L}$, but we won't treat them here.

2.3 Algorithms

In this section we will review some of the more common algorithms for optimizing modularity (and some of its alternatives). The problem of community detection is NP-hard in general [6], so that there is no (known[5]) efficient (polynomial time) algorithm for optimizing the objective function. The algorithms presented will thus be heuristics, and usually involve some stochasticity. This implies that it will not necessarily always find exactly the same partition. In fact, modularity often seems to have many near optimal partitions, making it difficult to obtain the global optimum, and the other methods are expected to show a similar degeneracy [19].

In order to test whether an algorithm is working correctly, and performs well, it is useful to construct test networks. These test networks—also known as benchmark networks—are constructed such that the community partition is known beforehand. Comparing the known partition to the partition detected by the algorithm provides evidence of how well the algorithm is performing. We will test some of the methods, and present their results. In spite of the NP-hardness of the problem, and that the algorithms are only heuristic, we will see they work reasonably well.

2.3.1 Simulated Annealing

Simulated Annealing (SA) is a general optimization technique [26]. The idea is that the search is allowed to explore a large part of the landscape at the beginning, but as the algorithm progresses, follows more and more the steepest descent trajectory (greedily) towards a (local) minimum. The basic idea is to analyse the difference in the objective function $\Delta\mathcal{H} = \mathcal{H}_{\text{after}} - \mathcal{H}_{\text{before}}$ when making a certain change to the partition. We will use $\Delta\mathcal{H} = \mathcal{H}_{\text{after}} - \mathcal{H}_{\text{before}}$ throughout this thesis, so that $\Delta\mathcal{H} < 0$ will always mean there is an improvement after some change, while $\Delta\mathcal{H} > 0$ indicates the prior situation was better (remember we are minimizing $\mathcal{H}$). Such a change can take many forms, but the changes usually considered are: moving a single node from one community to another; merging two communities; or splitting a community.

There are several choices available for accepting such a change. The idea is to also accept changes that worsen the partition (i.e. when $\Delta\mathcal{H} > 0$) with some probability that decreases as the algorithm progresses. The implementation from Reichardt and Bornholdt [41] works as follows. Consider moving node i from community c to d, and let the new communities be c' and d'. In terms of the community set we thus have that $C'_c = C_c \setminus i$ and $C'_d = C_d \cup i$. The change in the objective function is then

$$\Delta\mathcal{H}(\sigma_i = c \mapsto d) = (e_{id'} - \gamma_{\text{RB}}\langle e_{id'}\rangle) - (e_{ic'} - \gamma_{\text{RB}}\langle e_{ic'}\rangle) \tag{2.33}$$

---

[5] It is unlikely that any efficient algorithm will ever be found, part of the famous P = NP problem.

where $e_{ic'} = e_c - e_{c'} = \sum_j A_{ij}\delta(\sigma_i, c')$ is the number of edges from node i to community c' and $\langle e_{ic'}\rangle = \langle e_c\rangle - \langle e_{c'}\rangle = \sum_j p_{ij}\delta(\sigma_i, c')$ the expected number of edges from i to community c', and similarly so for d'. We consider all communities to which node i is connected, and the associated change in the objective function of $\Delta\mathcal{H}(\sigma_i = c \mapsto d)$. We then choose the new community with probability

$$\Pr(\sigma_i = d) = \frac{1}{Z}\exp\bigl[-\beta\Delta\mathcal{H}(\sigma_i = c \mapsto d)\bigr], \tag{2.34}$$

where $Z = \sum_d \exp\beta\Delta\mathcal{H}(\sigma_i = c \mapsto d)$ is the normalization factor. This is known as the Boltzmann probability distribution [24]. The parameter $\beta = 1/T$ is known as the inverse temperature. A high temperature (low β) gives nearly uniform probabilities, so that every change is chosen with almost equal probabilities. As the algorithm progresses, the temperature is lowered, for example after n changes, usually via $T' = \alpha T$ where $0 < \alpha < 1$ is some decay factor. Lower temperature leads to more narrow choices, and in the limit of $T \to 0$ only the moves with the maximum improvement of the objective function are chosen.

An alternative scheme was proposed by Guimerà et al. [20, 22]. Instead of considering all possible changes, we simply choose a random new community for a node. Similarly, a change can consist of merging two communities. Finally, a change can consist of splitting a community in two. All changes have a certain associated change in the objective function of $\Delta\mathcal{H}$ and the change is accepted with probability

$$\Pr(\text{accept change}) = \begin{cases} 1 & \text{if } \Delta\mathcal{H} < 0, \\ \exp(-\beta\Delta\mathcal{H}) & \text{if } \Delta\mathcal{H} \geq 0. \end{cases} \tag{2.35}$$

The change for moving a node i from community c to community d is already provided in Eq. (2.33). The change when merging two communities c and d into one new community c' is then

$$\Delta\mathcal{H}(\{c, d\} \mapsto c') = -e_{cd} + \gamma_{\text{RB}}\langle e_{cd}\rangle_{p_{ij}} \tag{2.36}$$

while the splitting of community c' into c and d is just the opposite

$$\Delta\mathcal{H}(c' \mapsto \{c, d\}) = -\Delta\mathcal{H}(\{c, d\} \mapsto c'), \tag{2.37}$$

with $e_{cd} = \sum_{ij} A_{ij}\delta(\sigma_i, c)\delta(\sigma_j, d)$ the number of edges between c and d and $\langle e_{cd}\rangle$ the expected number of such edges. A random split is unlikely to improve the partition, so some additional effort should be made to find a reasonably good candidate split, for example by using the eigenvector split (see Sect. 2.3.4), but we will not consider that here.

For both implementations the general idea remains the same. We consider a number of changes, which are accepted with a certain probability. After a certain number of changes, we lower the temperature, and repeat the procedure. When the objective function is no longer improved, the procedure terminates. The method moving only nodes is provided in Algorithm 1.

The exact calculations depend on the null model used. For the configuration null model, we have that $\langle e_c \rangle = K_c^2/2m$, and if we work out we obtain

$$\Delta\mathcal{H}(\sigma_i = c \mapsto d) = e_{id'} - e_{ic'} - \gamma_{\text{RB}}\frac{k_i}{m}(K_d - K_c + k_i) \tag{2.38a}$$

$$\Delta\mathcal{H}(\{c, d\} \mapsto c') = \gamma_{\text{RB}}\frac{K_c K_d}{2m} - e_{cd} \tag{2.38b}$$

$$\Delta\mathcal{H}(c' \mapsto \{c, d\}) = e_{cd} - \gamma_{\text{RB}}\frac{K_c K_d}{2m}, \tag{2.38c}$$

for respectively joining nodes, merging communities and splitting communities. For the ER null model, with $\langle e_c \rangle = pn_c^2$ where n_c is the size of community c, we obtain

$$\Delta\mathcal{H}(\sigma_i = c \mapsto d) = e_{id} - e_{ic} - \gamma_{\text{RB}}p((n_d + 1) - (n_c - 1)) \tag{2.39a}$$

$$\Delta\mathcal{H}(\{c, d\} \mapsto c') = \gamma_{\text{RB}}pn_c n_d - e_{cd} \tag{2.39b}$$

$$\Delta\mathcal{H}(c' \mapsto \{c, d\}) = e_{cd} - \gamma_{\text{RB}}pn_c n_d. \tag{2.39c}$$

Similar calculations can be derived for the other models.

Algorithm 1 Simulated Annealing (SA) method

```
function SA(Graph G)
  initialize σ_i ← i for all nodes i
  T ← some high number, β ← 1/T
  while improvement do
    for all nodes i do
      C_neigh ← {σ_j | (i, j) ∈ E} ∪ σ_i          ▷ Communities of neighbours
      for all communities d ∈ C_neigh do
        P_d ← exp(βΔH(σ_i = c ↦ d))
      end for
      σ_i ← RANDSAMPLE(P)                        ▷ Draw random community
    end for
    T ← α * T, β ← 1/T                            ▷ Lower temperature
  end while
  return σ
end function
```

2.3.2 *Greedy Improvement*

Graph partitioning itself is not new, and one heuristic method that has long been used, and which resembles the steps from Simulated Annealing (SA), is Kernighan-Lin (KL) improvement [25]. Although in the original formulation two nodes are swapped from their communities in order to keep the community sizes the same, this is not necessary for modularity optimization. So, the greedy improvement we consider here simply amounts to moving nodes from one community to another.[6] The difference with SA is that we choose greedily the best new community. In other words, the method loops (randomly) over all nodes, and determines for each node the community with the largest $\Delta\mathcal{H}$. It repeats these steps as long as there remain improvements.

More specifically, when considering node i we greedily check the increase in the objective function $\Delta\mathcal{H}(\sigma_i = c \mapsto d)$ if the node was moved from community c to d, as was already calculated in Eq. (2.33). Now instead of choosing the new community with a certain probability as defined in Eq. (2.34), we simply choose the community

$$s^* = \arg\max_s \Delta\mathcal{H}(\sigma_i = r \mapsto s) \tag{2.40}$$

which maximizes the change. This can be seen as the limit of the simulated annealing process for which $T \to 0$ (or $\beta \to \infty$). We consider all nodes (perhaps in random order), and repeat until no further improvement can be made.

Algorithm 2 Greedy method

```
function GREEDY(Graph G)
    initialize σ_i ← i for all nodes i
    while improvement do
        for all nodes i do
            C ← {σ_j | (i, j) ∈ E} ∪ σ_i                ▷ Communities of neighbours
            for all communities d ∈ C do
                Δ_d ← ΔH(σ_i = c ↦ d)
            end for
            σ_i ← arg max_d Δ_d                         ▷ Greedily, maximum choice
        end for
    end while
    return σ
end function
```

---

[6] There are some other greedy algorithms as well, for example [7, 8].

2.3.3 Louvain Method

The Louvain method for optimizing modularity [4] is one of the fastest and best algorithms available for optimizing modularity [29]. It makes changes to the partition similar to the greedy improvement, i.e. it always makes the optimal change at that moment. The trick that makes it so fast and yet work well, is that whenever no more changes can be made by moving nodes, we aggregate the graph, and rerun the same algorithm on the aggregated graph. This is then repeated until modularity can be no further increased.

Algorithm 3 Louvain method

```
function LOUVAIN(Graph G)
    σ ← GREEDY(G)                          ▷ Initial Greedy
    Σ ← σ                                  ▷ Use Σ for aggregate
    while improvement do
        G ← AGGREGATE(G, Σ)
        Σ ← GREEDY(G)                      ▷ Greedy on aggregate graph
        σ_i ← Σ_{σ_i} for all i            ▷ Correct σ according to Σ
    end while
    return σ
end function
```

The important detail is then of course that moving nodes in the aggregated graph should be equivalent to merging communities in the original graph. Hence, the aggregate method depends on the exact cost function used. Using the configuration null-model allows for a particularly straightforward aggregation. In that case, the new aggregated weighted adjacency matrix A' is constructed as follows

$$A'_{cd} = \sum_{ij} A_{ij}\delta(\sigma_i, c)\delta(\sigma_j, d) = e_{cd}$$

which simply creates a new node c for each community, and an edge to another new node community d has as weight the total number of edges between community c and d. The essential thing is now that joining two nodes in this graph A' should be equivalent to merging two communities in A. The benefit for joining nodes c and d in A' is

$$\Delta\mathcal{H} = -\left(A'_{cd} - \gamma_{\text{RB}}\frac{k'_c k'_d}{m}\right)$$

which is equivalent to joining communities c and d in A since $A'_{cd} = e_{cd} = \sum_{ij} A_{ij}\delta(\sigma_i, c)\delta(\sigma_j, d)$ the number of edges between communities c and d and $k'_c = \sum_d A'_{cd} = \sum_{ij} k_i\delta(\sigma_i, c)$ is the total degree in community c. Hence, joining two nodes is indeed equivalent to merging two communities as specified in Eq. (2.38b). This special feature of the configuration model (and modularity) allows this formulation to exploit this.

Algorithm 4 Aggregation for configuration null-model

function AGGREGATE(Graph G, Community σ)
  $A \leftarrow$ ADJACENCY(G)
  $A'_{cd} \leftarrow \sum_{ij} A_{ij}\delta(\sigma_i, c)\delta(\sigma_j, d)$
  return A'
end function

When using the ER null model this way of aggregating does not work correctly. Let us assume for an instance that we aggregated a graph according to this method. The benefit of merging node c and d in this aggregate graph, according to the ER null model is then

$$\Delta\mathcal{H} = A_{cc} + A_{dd} - A'_{cd} - \gamma_{\text{RB}}p$$

while this should actually be

$$\Delta\mathcal{H} = A_{cc} + A_{dd} - A'_{cd} - \gamma_{\text{RB}}pn_cn_d$$

where n_c and n_d are the number of nodes in community c and d. Using this method of aggregating then clearly does not work.

In order to make this step of aggregating the graph work for the ER null-model we need to introduce the node size. In the aggregate graph, the node size will then represent the number of nodes in the community (i.e. the community size). So, for the initial graph we set the node size to $n_i = 1$ for all nodes, and upon aggregating we will set the node size $n_c = \sum_i n_i\delta(\sigma_i, c)$ of community c, i.e. the new node in the aggregated graph, equal to the sum of the node sizes within the community.

Notice that we can use the same type of aggregation for CPM (and by extension RN). Since we can also apply the greedy algorithm to CPM, the Louvain method is easily applied to CPM as well.

Algorithm 5 Aggregation for ER null-model & CPM

function AGGREGATE(Graph G, Community σ)
  $A \leftarrow$ ADJACENCY(G)
  $A'_{cd} \leftarrow \sum_{ij} A_{ij}\delta(\sigma_i, c)\delta(\sigma_j, d)$
  $n_c \leftarrow \sum_i n_i\delta(\sigma_i, c)$
  return A', n'
end function

2.3.4 Eigenvector

We can also take a matrix analysis perspective [38]. If we define the modularity matrix B with entries

$$B_{ij} = a_{ij} A_{ij} - b_{ij}(1 - A_{ij}) \tag{2.41}$$

and S the $n \times q$ community matrix, such that $S_{ic} = 1$ if node i is in community c and 0 otherwise, we can write our objective function as

$$\mathcal{H} = -\sum_{ij} \sum_{c} B_{ij} S_{ic} S_{jc} = -\text{Tr} S^\top B S, \tag{2.42}$$

since $S_{ic} S_{jc} = 1$ if $\sigma_i = \sigma_j = c$ and 0 otherwise, so that $\sum_c S_{ic} S_{jc} = \delta(\sigma_i, \sigma_j)$, and $\sum_i S_{ic} S_{id} = 0$ for $c \neq d$. Here $S^\top$ denotes the transpose of S (i.e. $S^\top_{ij} = S_{ji}$). Since each node should be in exactly one community, we have the constraint that $S_{ic} \in \{0, 1\}$ and $\sum_c S_{ic} = 1$. From this it also follows that $\text{Tr} S^\top S = n$ and that the columns of S are mutually orthogonal. For undirected graphs B is symmetric (i.e. $B = B^\top$), and we can decompose $B = U \Lambda U^\top$ where Λ is a diagonal vector containing the eigenvalues $\lambda_1 \geq \lambda_2 \geq \cdots \geq \lambda_n$ with U an orthogonal matrix (i.e. $U U^\top = I_n$ is the identity matrix) containing the associated eigenvectors. Plugging this in leads to

$$\begin{aligned} \mathcal{H} &= -\text{Tr} S^\top U \Lambda U^\top S \\ &= -\text{Tr} \Lambda U^\top S S^\top U, \end{aligned}$$

So, for all $\lambda_i > 0$ we should put as much weight as possible in $U_i^\top S S^\top U_i$. Without the constraint that $S_{ic} \in \{0, 1\}$ this would be simply optimized by taking the column S_i proportional to u_i for $\lambda_i > 0$, and the rest 0. Because of the constraints that $S_{ic} \in \{0, 1\}$ this is not straightforward, and usually only a partitioning in two groups is considered. This is known as (recursive) spectral bisectioning. The basic idea is to recursively split communities, until we can no longer divide the sub parts.

For spectral bisectioning, it is simpler to use a single vector s to indicate two groups as $s_i = -1$ if i is in group 1 and $s_i = 1$ if i is in group 2. Then $\frac{1}{2}(s_i s_j + 1) = \delta(\sigma_i, \sigma_j)$, and we can write

$$\mathcal{H} = -\sum_{ij} B_{ij} \frac{1}{2}(s_i s_j + 1)$$

which is up to a multiplicative and additive constant equivalent to

$$\mathcal{H} = -s^\top B s, \tag{2.43}$$

with $s^\top s = n$. If we relax the problem by allowing s to take on real values, $s^\top B s$ is similar to a Rayleigh quotient, for which it is well known that it is maximized by taking s proportional to u where u is the eigenvector associated to λ_1 the largest eigenvalue of B. Hence, if we take

$$s_i = \begin{cases} 1 & \text{if } u_i \geq 0, \\ -1 & \text{if } u_i < 0, \end{cases}$$

this is the vector s with $s_i \in \{-1, +1\}$ for which $\|s - u\|$ is minimal.

We can then recursively apply this method to a single community. Let B^c be the $n_c \times n_c$ submatrix of B corresponding to community c. The improvement of $\mathcal{H}$ by dividing community c in two, again denoted by the vector $s \in \{-1, +1\}^{n_c}$, can then be described by

$$\Delta\mathcal{H} = -\sum_{ij} B^c_{ij} \frac{1}{2}(s_i s_j + 1) - B^c_{ij}$$

which by removing parts that don't depend on the optimization reduces to

$$\Delta\mathcal{H} = -\sum_{ij} B^c_{ij} s_i s_j = -s^\top B^c s \tag{2.44}$$

similar as before. So, we follow the same procedure. However, we must ensure that the total contribution is positive still, so that $\Delta\mathcal{H}$ in Eq. (2.44) must obey

$$\Delta\mathcal{H} = -s^\top B^c s < -e^\top B^c e$$

with $e = (1, \ldots, 1)$ the vector of all ones. In other words, as long as subdividing puts more weight within the subdivided community as there is in total within the community, we should continue splitting. Notice that this is similar to the condition that $\Delta\mathcal{H}(c' \mapsto \{c, d\}) > 0$ for splitting community c' into community c and d in Eq. (2.37). Furthermore, notice that for the RB model with $\gamma_{\text{RB}} = 1$ we have that $e^\top Be = 0$ by definition of modularity, so that we can use the same condition.

Algorithm 6 Recursive eigenvector bisection

function EIGENVEC(Modularity matrix B)
  $u \leftarrow$ largest eigenvector of B
  $\sigma_i \leftarrow \begin{cases} -1 & \text{if } u_i \geq 0, \\ 1 & \text{if } u_i < 0. \end{cases}$
  if $\sigma^\top B\sigma > e^\top Be$ **then** ▷ If improvement
    $\Sigma_1 \leftarrow$ EIGENVEC($B(\sigma = -1, \sigma = -1)$) ▷ Submatrix for $\sigma_i = -1$
    $\Sigma_2 \leftarrow$ EIGENVEC($B(\sigma = 1, \sigma = 1)$) ▷ Submatrix for $\sigma_i = 1$
    $\sigma \leftarrow$ Combine Σ_1 and Σ_2
  else
    $\sigma_i \leftarrow 1$ ▷ Otherwise, don't split
  end if
  return σ
end function

2.4 Benchmarks

In order to know whether these algorithms and methods work effectively, we now turn to methods for testing them. This involves two parts. First we have to construct good test networks with some planted partition, so that we can check if some community detection method is able to uncover this planted partition. Secondly, we need some measure to compare the computed partition to the planted partition. Finally, we will provide some results comparing different methods.

2.4.1 Test Networks

One of the first problems in generating test networks is that there is no definitely agreed upon definition of a community. However, as stated earlier, there is some consensus on some common features: the communities should be relatively dense, and relatively well separated from the rest of the network. Although specific details might not be agreed upon exactly, this often is the foundation upon which test networks are constructed. Still, we should keep in mind that different definitions of communities or good partitions might yield a partition different from the planted partition. This does not necessarily imply the method does not work correctly, because the definition of community simply differs. Nonetheless, if some method is unable to detect correctly the planted partition whereas other methods do, it does indicate it might not be the appropriate method for these type of test networks.

The first to propose such test networks were [18], and remained the common benchmark for some time [10]. In general, test networks are constructed as follows. We wish to build a network of q communities of each n_c nodes with average degree $\langle k \rangle$. The total number of nodes is then $n = qn_c$ and the total number of edges $m = \langle k \rangle n/2$. Furthermore, we would like to control the difficulty of detecting communities. The denser communities are, and the better separated from the rest of the network, the easier it is to detect such communities. Hence, we will introduce a mixing parameter $0 \leq \mu \leq 1$ such that each node will have about $(1 - \mu)\langle k \rangle$ edges within its community, and about $\mu\langle k \rangle$ edges outside its community. Such a network can be easily constructed as follows. We pick a random node i and with probability μ we will link to a node outside of its community, and with probability $1 - \mu$ we link to a node within its community. We will add in total $\langle k \rangle n/2$ edges. Easily partitioned networks are constructed using a low μ and this gets progressively more difficult for higher μ. The common test setting introduced by Mark Newman used $q = 4$ communities of $n_c = 32$ nodes each, with μ varying from 0 to 1.

One question concerns until what point μ we expect communities to exist. A reasonable limit is that the average density within a community should be higher than the average density between communities. Beyond this threshold communities become very fuzzy (regardless of the definition) and are unlikely to be detected by any method.

Let us first calculate the inner density for a community of size n_c. Each of the n_c nodes has on average $(1-\mu)\langle k\rangle$ edges within its community, and the density is therefore

$$p_{\text{in}} = \frac{(1-\mu)\langle k\rangle}{n_c - 1}. \tag{2.45}$$

The rest of the $\mu\langle k\rangle$ edges per node will be distributed across the rest of the network. Since these edges get distributed over $n - n_c$ nodes, they will be more dispersed in general. The average density is then simply

$$p_{\text{out}} = \frac{\mu\langle k\rangle}{n - n_c}. \tag{2.46}$$

A community of n_c nodes in the test network is then well-defined as long as $p_{\text{in}} > p_{\text{out}}$, which yields

$$\mu < \frac{n - n_c}{n - 1} \approx \frac{q-1}{q}. \tag{2.47}$$

In other words, the probability for a link within a community μ should be smaller than the proportion of nodes outside the community. Notice this is independent of the total size of the network, the average degree, and the size of the communities, and depends only on the number of communities q (up to a correction term of $\frac{1}{n_c}$). For the regular test setting of $q = 4$ communities this yields $\mu < 0.75$, contrary to what was believed earlier that the communities would be defined up to $\mu = 0.5$.

In fact, such a test network most closely resembles a random network around $\mu \approx (q-1)/q$. For smaller μ the network exhibits a community structure. For higher μ however, the network still has a very particular structure. In that case, there are few links within communities, and many between communities. In other words, it starts to show a multi-partite structure.

Although such a test network is fine, it is far from realistic. Most networks show a skewed degree distribution with a fat-tail. They have many nodes with a low degree, and some nodes with an extremely high degree. The above test networks on the other hand have a Poissonian degree distribution, such that most of the nodes have about the same degree $k_i \approx \langle k\rangle$. Most empirical results of community detection suggests the community sizes are also highly skewed, while in these test networks each community is of exactly the same size. This could lead to a potential bias when benchmarking methods, since it only looks to whether a method can find communities in this particular test setting. In order to overcome these issues it was suggested to create test networks that have a power-law degree and community size distribution by Lancichinetti et al. [30], now commonly known as the LFR benchmark. Additionally, weights of links can be introduced, which realistically should also take a power-law distribution. These weights can again be distributed differently within and between communities.

Furthermore, many complex networks show some form of hierarchical structure [30]. In order to test for this, hierarchical test networks would be needed. So, instead of only having a single partition in communities, each community at the lowest level is embedded in increasingly larger communities. Instead of specifying then a single μ for the probability of having links outside the community, we specify $\mu_1, \mu_2, \ldots, \mu_l$ for l different levels, with each level i being embedded in the $i-1$ level. Level 1 is then the coarsest, highest level, and l the lowest most refined level. Of course, these probabilities are limited to $\sum_l \mu_l < 1$.

The limits of the densities remain rather similar, but now depend on the level we are looking at. Let us take a look to a two level hierarchy. The corresponding densities then are

$$\begin{aligned} p_1^{in} &= (1-\mu_1)\langle k\rangle/(n_{c,1}-1) \\ p_1^{out} &= \mu_1\langle k\rangle/(n-n_{c,1}) \\ p_2^{in} &= (1-\mu_1-\mu_2)\langle k\rangle/(n_{c,2}-1) \\ p_2^{out} &= (\mu_1+\mu_2)\langle k\rangle/(n-n_{c,2}) \end{aligned}$$

where $n_{c,1}$ is the community size at level 1 and $n_{c,2}$ the community size at level 2. The second level then remains detectable until

$$\mu_1 + \mu_2 < \frac{n-n_{c,2}}{n-1}.$$

Similarly, the first level is well defined until

$$\mu_1 < \frac{n-n_{c,1}}{n-1}.$$

Both limits are similar to the original limit in Eq. (2.47) but, there is a trade-off between the fine (μ_2) and course level (μ_1). Whenever the coarse level is less well defined, the corresponding limit for the finer level becomes smaller.

2.4.2 Comparing Partitions

Once a test network with a known partition is available, we need a measure for stating how well a certain method is able to recover this known partition. Various measures are suitable for this, but two of the most common ones are the normalized mutual information (NMI) and the variation of information (VI). The NMI measures how much information we have about one partition knowing the other. The VI is a true metric, and is closely related to the NMI. Benchmark results are usually provided in NMI, but VI seems somewhat more sensitive to small deviations.

Both measures have their origins in information theory, of which the basics have been provided in Sect. 2.2.7 (see pp. 23–26). The mutual information is defined as

$$\begin{aligned} I(X, Y) &= H(X) - H(X \mid Y) = H(Y) - H(Y \mid X) \\ &= H(X) + H(Y) - H(X, Y). \end{aligned}$$

Hence, if X and Y are two independent variables, $H(X, Y) = H(X) + H(Y)$ and $I(X, Y) = 0$. On the other hand, if X is completely determined by Y then $H(X, Y) = H(X) = H(Y)$ and $I(X, Y) = I(X, X) = H(X)$. Hence, we can normalize $I(X, Y)$ by $H(X) + H(Y)$ and arrive at at the normalized mutual information

$$\text{NMI}(X, Y) = \frac{2I(X, Y)}{H(X) + H(Y)}, \tag{2.48}$$

which is always $0 \leq \text{NMI}(X, Y) \leq 1$. The Variation of Information (VI) can then be defined as

$$\text{VI}(X, Y) = H(X) + H(Y) - 2I(X, Y), \tag{2.49}$$

$$= 2H(X, Y) - H(X) - H(Y), \tag{2.50}$$

Since $I(X, Y) = H(X)$ if and only if X is completely determined by Y then $\text{VI}(X, X) = 0$. Otherwise, since $2I(X, Y) \leq H(X) + H(Y)$, we have that $\text{VI}(X, Y) \geq 0$. Furthermore, notice that $\text{VI}(X, Z) \leq \text{VI}(X, Y) + \text{VI}(Y, Z)$, since the inequality

$$\begin{aligned} 2H(X, Z) - H(X) - H(Z) \leq\, & 2H(X, Y) + 2H(Y, Z) \\ & - H(X) - 2H(Y) - H(Z) \end{aligned}$$

is equivalent to

$$\begin{aligned} H(X, Z) &\leq H(X, Y) + H(Y, Z) - H(Y) \\ H(X \mid Z) &\leq H(X \mid Y) + H(Y \mid Z). \end{aligned}$$

The last inequality holds because

$$\begin{aligned} & H(X \mid Y) + H(Y \mid Z) - H(X \mid Z) \\ & \quad \geq H(X \mid Y, Z) + H(Y \mid Z) - H(X \mid Z) \\ & \quad = H(X, Y \mid Z) - H(X \mid Z) \geq 0 \end{aligned}$$

In other words, the $\text{VI}(X, Y)$ is a true metric, and can be interpreted to provide a distance between the random variables X and Y. There are several ways to normalize this quantity, for example by dividing by $I(X, Y)$ or by $\max\{H(X), H(Y)\}$, but this is not often considered [29, 34].

When it comes to comparing partitions, these quantities are used as follows. Let C and D be two partitions, such that there are n_c nodes in community c in C, n_d nodes in community d in D and n_{cd} nodes that are in community c in C and in community

d in D. The probability a random node is in community c is then $p_c = n_c/n$, and likewise we can define the probability $p_{cd} = n_{cd}/n$. Working this out for mutual information, we thus arrive at

$$I(C, D) = -\sum_{cd} \frac{n_{cd}}{n} \log \left(n \frac{n_{cd}}{n_c n_d} \right)$$

and

$$H(C) = -\sum_{c} \frac{n_c}{n} \log \frac{n_c}{n}.$$

The other quantities follow readily. The baseline is that NMI $= 1$ (and so VI $= 0$) whenever $C = D$ the two partitions are equal. So when comparing a method to the known partition, if a method works well, NMI ~ 1, and VI ~ 0.

Other well known measures for comparing partitions are the (adjusted) rand index and Jaccard index [15, 47, 51]. This is based on checking how many pairs of nodes are clustered in the same manner. The number of pairs of nodes that are clustered in the same way in both partitions can be obtained as

$$a = \sum_{cd} n_{cd}, \tag{2.51}$$

where n_{cd} denotes the number of nodes that are in community c in partition C and in community d in partition D. The number of pairs of nodes that are clustered both in different communities—so the number of pairs of nodes i and j such that they are not in the same community in partition C and neither in partition D can be described by

$$b = \binom{n}{2} + \sum_{cd} n_{cd} - \sum_{c} n_c^C - \sum_{c} n_c^D \tag{2.52}$$

where n_c^C refers to the number of nodes in community c in partition C. Then the rand index is defined as

$$\mathrm{RI}(C, D) = \frac{a + b}{\binom{n}{2}}, \tag{2.53}$$

namely the fraction of pairs of nodes that are classified in the same manner (belonging both to the same community is both partitions are both to different communities in both partitions). This measure varies between 0 and 1 with 1 indicating two identical partitions C and D while 0 indicates two completely different partitions. There exists an adjusted version which takes into account the fact that the rand index for two random partition already attains some similarity. The Jaccard index is defined as

$$\mathrm{J}(C, D) = \frac{a}{\binom{n}{2} - b}. \tag{2.54}$$

Compared to the VI both measures have some drawbacks [34], although no measure is perfectly fit for all situations. For benchmarks in community detection however, the NMI has become the standard, although the rand index, Jaccard index and other variants are used in other domains.

2.4.3 Results

Not all models work equally well. We have tested extensively the RB model using the configuration null model and the ER null model, CPM and Infomap. For the RB model the "natural" parameter is $\gamma_{\text{RB}} = 1$, which then corresponds to modularity for the configuration model. For Infomap there is no parameter present, so there is little to choose there. For CPM there is no such "natural" parameter, and one would have to look which γ_{CPM} works best (we will touch upon this issue in Sect. 4.1). However, given that we know how we generate the benchmark networks, we can calculate the optimal parameter γ^*_{CPM} for uncovering the planted partition. Since the CPM model and the RB model are equal for the ER null model when using $\gamma_{\text{CPM}} = \gamma_{\text{RB}} p$, this also corresponds to the optimal parameter for the RB model with the ER null model. For the configuration null model we can choose a similar optimal parameter value, in order to detect the planted partition as well as possible.

Let us calculate this optimal parameter value. We denote by p_{in} the average density within a community, and by p_{out} the average density between a community and the rest of the network. For CPM to correctly detect these communities we should set $\gamma_{\text{CPM}} > p_{\text{in}}$ so that it doesn't split communities of that density, while $\gamma_{\text{CPM}} < p_{\text{out}}$ so that it doesn't merge communities either. We have already calculated these densities before in Eqs. (2.45) and (2.46), and we set

$$\gamma^*_{\text{CPM}} = \gamma^*_{\text{RB}} p = \frac{\langle p_{\text{in}} \rangle + \langle p_{\text{out}} \rangle}{2}$$

where $\langle p_{\text{in}} \rangle$ indicates we have taken the average p_{in} over all community sizes.

In order to calculate a similar optimal resolution parameter for the configuration model, notice that we should have that the inner "degree density" $\tilde{p}_{\text{in}} = \frac{e_c}{\langle e_c \rangle_{\text{conf}}}$ should be lower than γ_{RB}, while the outer "degree density" should be higher than γ_{RB}. The number of edges within a community is simply $e_c = n_c \langle k \rangle (1 - \mu)$, and the expected sum of degrees $K_c = n_c \langle k \rangle$. Furthermore, the total number of expected edges is $2m = n \langle k \rangle$, so that we obtain

$$\begin{aligned} \tilde{p}_{\text{in}} &= \frac{e_c}{\langle e_c \rangle_{\text{conf}}} \\ &= \frac{n_c \langle k \rangle (1 - \mu)}{\frac{(n_c \langle k \rangle)^2}{n \langle k \rangle}} \\ &= \frac{n(1 - \mu)}{n_c}. \end{aligned}$$

The outer "degree density" can be similarly calculated. The number of external edges remains $e_{c*} = n_c\mu\langle k\rangle$ as before (where the $*$ denote the rest of the network). The expected number of edges is $\langle e_{c*}\rangle = K_c K_*/2m$, and so becomes $\langle e_{c*}\rangle = n_c(n-n_c)\langle k\rangle^2/2m$, so that the outer "degree density" is

$$\begin{aligned}\tilde{p}_{\text{out}} &= \frac{e_{c*}}{\langle e_{c*}\rangle_{\text{conf}}} \\ &= \frac{n_c\mu\langle k\rangle}{\frac{n_c(n-n_c)\langle k\rangle^2}{n\langle k\rangle}} \\ &= \frac{\mu n}{n-n_c}.\end{aligned}$$

Similar as before, we set the RB resolution parameter for the configuration model at

$$\gamma^*_{\text{RB}} = \frac{\langle\tilde{p}_{\text{in}}\rangle + \langle\tilde{p}_{\text{out}}\rangle}{2}$$

Notice that we can do a similar analysis as before, trying to calculate the point at which communities are no longer well defined, but use the "degree densities" to do so. Working out the inequality $\tilde{p}_{\text{in}} > \tilde{p}_{\text{out}}$ we obtain that up until

$$\mu < \frac{n-n_c}{n} \approx \frac{q-1}{q}$$

the communities are well defined. Hence, this does not change anything in comparison to our earlier analysis in Eq. (2.47).

The results for the different methods are displayed in Fig. 2.1. On the y-axis it shows the NMI as defined earlier, while on the x-axis the mixing parameter μ is shown. For each value of the mixing parameter μ we generate 100 LFR benchmark networks. We have used the Louvain algorithm for all models, since earlier analysis showed the Louvain algorithm works at least as well as many other algorithms, but is much faster. For a more extensive comparison between different algorithms, refer to Lancichinetti and Fortunato [29].

It can be clearly seen that CPM performs well. The difference in performance of the CPM model in comparison to the RB model using the ER null model is especially striking. Obviously then, setting $\gamma_{\text{CPM}} = p$ is in general not a very good strategy, and for general networks one should carefully analyse at which resolution the network contains meaningful partitions, a topic we will review briefly in Sect. 4.1.

A similar effect also shows for modularity (or the RB model using the configuration model), such that when γ_{RB} is chosen appropriately (i.e. using $\gamma_{\text{RB}} = \gamma^*_{\text{RB}}$) the method will perform better than at the ordinary resolution $\gamma_{\text{RB}} = 1$. Indeed, the results of the CPM model and the RB model using the configuration null model using γ_{RB} are rather comparable, although the latter's performance drops less quickly, and then outperforms CPM. Interestingly, when we use the ordinary resolution $\gamma_{\text{RB}} = 1$, it becomes more difficult to detect communities in large networks using the

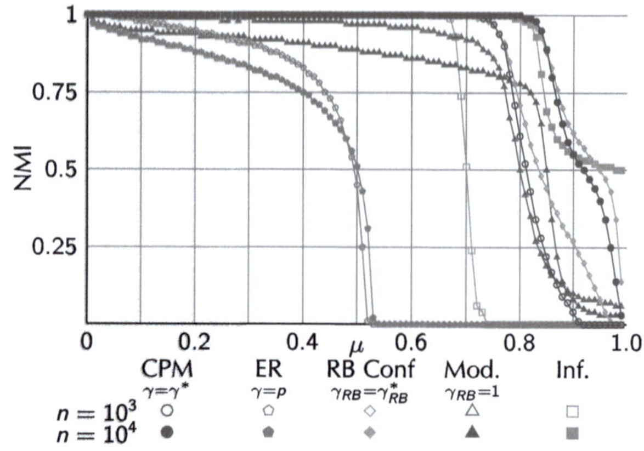

Fig. 2.1 Benchmark results

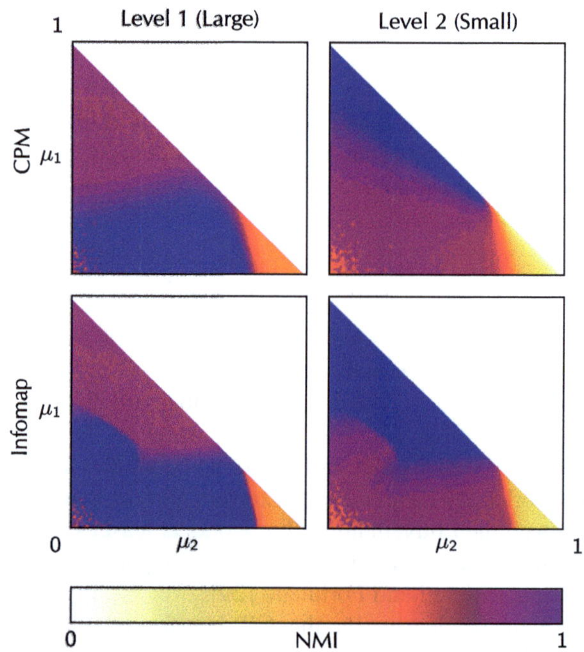

Fig. 2.2 Hierarchical benchmark results

configuration model. This contrasts with the results when we choose the appropriate resolution parameter γ^*_{CPM}, γ^*_{RB} and indeed also for the Infomap method. Indeed the communities should become more clearly discernible for larger networks when the community sizes remain similar. The limit of community detection as calculated earlier is about $\mu^* = \frac{q-1}{q} \approx 0.92$ for $n = 10^3$ and $\mu^* \approx 0.99$ for $n = 10^4$. The models with the tuned resolution parameters work quite well and approach this upper limit to some extent. Surprisingly, both methods outperform the Infomap method, which performed superbly in previous tests [29], when the appropriate resolution parameter is chosen.

We have also performed extensive tests on hierarchical networks, where the method also performs well, and is able to extract the two different levels of communities effectively, as displayed in Fig. 2.2. For relatively low $\mu_2 \lessapprox 0.7$, the first (larger) level becomes more clear for low μ_1, while the second (smaller) level becomes more clear for larger μ_1. This is both the case for a recent hierarchical version of the Infomap method [45] and the CPM method. The Infomap method seems to be slightly better at detecting the planted communities, but the CPM method remains highly competitive. The possibility for having various scales of description of the network seems important, as many networks seem to have at least some hierarchical structure.

References

1. Aldecoa R, Marín I (2013) Surprise maximization reveals the community structure of complex networks. Sci Rep 3:1060. doi:10.1038/srep01060
2. Arenas A, Fernandez A, Gomez S (2007) Analysis of the structure of complex networks at different resolution levels. New J Phys 10(5):23. doi:10.1088/1367-2630/10/5/053039. arXiv:physics/0703218
3. Bichot CE, Siarry P (2011) Graph partitioning. Wiley, New York. ISBN 184821233X
4. Blondel VD, Guillaume JL, Lambiotte R, Lefebvre E (2008) Fast unfolding of communities in large networks. J Stat Mech Theory Exp 2008(10):P10008. doi:10.1088/1742-5468/2008/10/P10008
5. Bollobás B (2001) Random graphs, 2nd edn. Cambridge University Press, Cambrige
6. Brandes U, Delling D, Gaertler M, Goerke R, Hoefer M et al (2006) Maximizing modularity is hard. arXiv:physics 0608255. arXiv:physics/0608255
7. Brandes U, Delling D, Gaertler M, Gorke R, Hoefer M et al (2008) On modularity clustering. IEEE Trans Knowl Data Eng 20(2):172–188. doi:10.1109/TKDE.2007.190689
8. Clauset A, Newman M, Moore C (2004) Finding community structure in very large networks. Phys Rev E 70(6):1–6. doi:10.1103/PhysRevE.70.066111
9. Cover, TM and Thomas, JA (2012). Elements of Information Theory, vol 2012. Wiley, New York. ISBN 1118585771
10. Danon L, Díaz-Guilera A, Duch J, Arenas A (2005) Comparing community structure identification. J Stat Mech Theory Exp 2005:P09008. doi:10.1088/1742-5468/2005/09/P09008
11. Delvenne JC, Yaliraki SN, Barahona M (2010) Stability of graph communities across time scales. Proc Natl Acad Sci USA 107(29):12755–12760. doi:10.1073/pnas.0903215107
12. Diestel R (2010) Graph theory, 4h edn. Springer, Berlin. ISBN 978-3-642-14278-9
13. Doreian P, Batagelj V, Ferligoj A (2005) Generalized blockmodeling. Cambridge University Press, Cambridge. ISBN 9780521840859
14. Duch J, Arenas A (2005) Community detection in complex networks using extremal optimization. Phys Rev E 72(2):1–4. doi:10.1103/PhysRevE.72.027104
15. Everitt BS, Landau S, Leese M (2001) Cluster analysis. Wiley, New York. ISBN 9780340761199
16. Fortunato S (2010) Community detection in graphs. Phys Rep 486(3–5):75–174. doi:10.1016/j.physrep.2009.11.002
17. Fouss F, Pirotte A, Renders JM, Saerens M (2007) Random-walk computation of similarities between nodes of a graph with application to collaborative recommendation. IEEE Trans Knowl Data Eng 19(3):355–369. doi:10.1109/TKDE.2007.46

18. Girvan M, Newman MEJ (2002) Community structure in social and biological networks. Proc Natl Acad Sci USA 99(12):7821–7826. doi:10.1073/pnas.122653799
19. Good BH, de Montjoye YA, Clauset A (2010) Performance of modularity maximization in practical contexts. Phys Rev E 81(4):046106. doi:10.1103/PhysRevE.81.046106
20. Guimerà R, Mossa S, Turtschi A, Amaral LAN (2005) The worldwide air transportation network: anomalous centrality, community structure, and cities' global roles. Proc Natl Acad Sci USA 102(22):7794–7799. doi:10.1073/pnas.0407994102
21. Guimerà R, Nunes Amaral LA (2005) Functional cartography of complex metabolic networks. Nature 433(7028):895–900. doi:10.1038/nature03288
22. Guimerà R, Sales-Pardo M, Amaral L (2004) Modularity from fluctuations in random graphs and complex networks. Phys Rev E 70(2):025101. doi:10.1103/PhysRevE.70.025101
23. Jain A, Dubes R (1988) Algorithms for clustering data. Prentice-Hall, Englewood Cliffs. ISBN 978-0130222787
24. Jaynes E (1957) Information theory and statistical mechanics. Phys Rev 106(4):620–630. doi:10.1103/PhysRev.106.620
25. Kernighan BW, Lin S (1970) An efficient heuristic procedure for partitioning graphs. Bell Syst Tech J 49(1):291–307
26. Kirkpatrick S, Gelatt CD, Vecchi MP (1983) Optimization by simulated annealing. Science (NY) 220(4598):671–680. doi:10.1126/science.220.4598.671
27. Kolaczyk ED (2009) Statistical analysis of network data: methods and models. Springer, Berlin. ISBN 9780387881461
28. Lambiotte R, Delvenne JC, Barahona M (2008) Laplacian dynamics and multiscale modular structure in, networks, pp 1–29. arXiv:0812.1770
29. Lancichinetti A, Fortunato S (2009) Community detection algorithms: a comparative analysis. Phys Rev E 80(5):056117. doi:10.1103/PhysRevE.80.056117
30. Lancichinetti A, Fortunato S, Radicchi F (2008) Benchmark graphs for testing community detection algorithms. Phys Rev E 78(4):46110. doi:10.1103/PhysRevE.78.046110
31. Lancichinetti A, Radicchi F, Ramasco JJ, Fortunato S (2011) Finding statistically significant communities in networks. PloS ONE 6(4):e18961. doi:10.1371/journal.pone.0018961
32. Leicht E, Newman M (2008) Community structure in directed networks. Phys Rev Lett 100(11):1–4. doi:10.1103/PhysRevLett.100.118703
33. MacKay D (2003) Information theory, inference and learning algorithms. Cambridge University Press, Cambridge. ISBN 9780521642989
34. Meilă M (2007) Comparing clusterings-an information based distance. J Multivar Anal 98(5):873–895. doi:10.1016/j.jmva.2006.11.013
35. Newman M (2004) Fast algorithm for detecting community structure in networks. Phys Rev E 69(6). doi:10.1103/PhysRevE.69.066133
36. Newman M (2010) Networks: an introduction. Oxford University Press, Oxford. ISBN 0199206651
37. Newman M, Girvan M (2004) Finding and evaluating community structure in networks. Phys Rev E 69(2):026113. doi:10.1103/PhysRevE.69.026113
38. Newman MEJ (2006) Finding community structure in networks using the eigenvectors of matrices. Phys Rev E 74(3):036104+. doi:10.1103/PhysRevE.74.036104
39. Porter MA, Onnela JP, Mucha PJ (2009) Communities in networks. Not AMS 56(9):1082–1097. arXiv:0902.3788
40. Raghavan U, Albert R, Kumara S (2007) Near linear time algorithm to detect community structures in large-scale networks. Phys Rev E 76(3):036106. doi:10.1103/PhysRevE.76.036106
41. Reichardt J, Bornholdt S (2006) Statistical mechanics of community detection. Phys Rev E 74(1):016110+. doi:10.1103/PhysRevE.74.016110
42. Reichardt J, White DR (2007) Role models for complex networks. Eur Phys J B 60(2):217–224. doi:10.1140/epjb/e2007-00340-y
43. Ronhovde P, Nussinov Z (2010) Local resolution-limit-free Potts model for community detection. Phys Rev E 81(4):046114. doi:10.1103/PhysRevE.81.046114. arXiv:0803.2548v4

44. Rosvall M, Bergstrom CT (2008) Maps of random walks on complex networks reveal community structure. Proc Natl Acad SciUSA 105(4):1118–1123. doi:10.1073/pnas.0706851105
45. Rosvall M, Bergstrom CT (2011) Multilevel compression of random walks on networks reveals hierarchical organization in large integrated systems. PloS ONE 6(4):e18209. doi:10.1371/journal.pone.0018209. arXiv:1010.0431
46. Schaeffer SE (2007) Graph clustering. Comput Sci Rev 1(1):27–64. doi:10.1016/j.cosrev.2007.05.001
47. Theodoridis S, Koutroumbas K (2006) Pattern recognition. Academic Press, New York. ISBN 9780080513614
48. Tibély G, Kertész J (2008) On the equivalence of the label propagation method of community detection and a Potts model approach. Phys A Stat Mech Appl 387(19–20):4982–4984. doi:10.1016/j.physa.2008.04.024
49. Traag VA, Van Dooren P, Nesterov Y (2011) Narrow scope for resolution-limit-free community detection. Phys Rev E 84(1):016114. doi:10.1103/PhysRevE.84.016114. arXiv:1104.3083
50. Wasserman S, Faust K (1994) Social network analysis. Cambridge University Press, Cambridge
51. Xu R, Wunsch D (2008) Clustering. Wiley, New Jersey. ISBN 9780470382783
52. Yen L, Fouss F, Decaestecker C, Francq P, Saerens M (2009) Graph nodes clustering with the sigmoid commute-time kernel: a comparative study. Data Knowl Eng 68(3):338–361. doi:10.1016/j.datak.2008.10.006

Chapter 3
Scale Invariant Community Detection

Modularity has been intensively studied the past decade, and although there are some positive aspects to it, it also has some problematic issues. One of the biggest advantages of modularity compared to older clustering methods is that it is not necessary to specify the number of clusters beforehand. Rather, the number of communities emerges naturally from the network at hand. However, this also seems to bring some issues along with it. We will discuss these issues here. In the following section we will focus specifically on modularity, Eq. (2.11). Most of the other introduced models address in some way or the other some of the issues discussed here, and we will analyse them after modularity. We will then define the problem more formally, and investigate what models there might be that are able to evade these problems.

3.1 Issues with Modularity

In this section, we will focus exclusively on modularity, and see some of its problems. Although the method has several problems, the most important one is that of the resolution limit, which we will now discuss first.

3.1.1 Resolution Limit

The most famous drawback of modularity is that of the so-called resolution limit. The problem is that some small communities in larger graphs cannot be detected by modularity. This is problematic, because the communities detected by modularity should then be split to uncover the "true" communities. This problem is usually studied by analysing a ring of cliques, and was actually introduced through this example.

V. Traag, *Algorithms and Dynamical Models for Communities and Reputation in Social Networks,* Springer Theses, DOI: 10.1007/978-3-319-06391-1_3,

Fig. 3.1 Ring of cliques

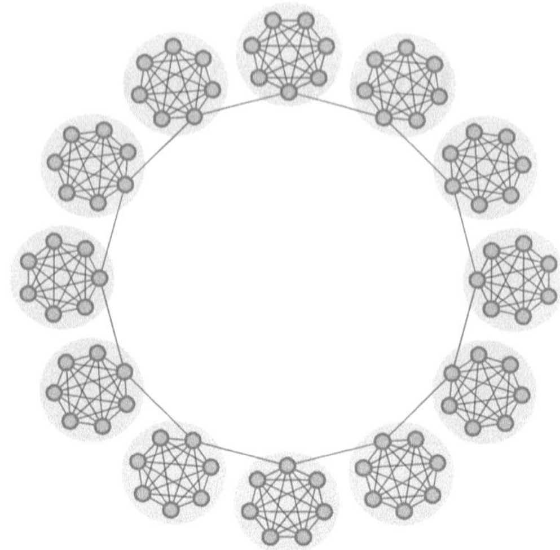

The ring of cliques is a graph that consists of r cliques (complete subgraphs) each of n_c nodes, with only a single link between each clique, and is displayed in Fig. 3.1. The graphs thus contains r densest possible subgraphs which are as sparsely connected as possible (only a single link). Intuitively each of these cliques should thus represent a community. In fact, this is the most modular (connected) network possible. Yet, modularity may counterintuitively merge these cliques.

In order to see this, let us first calculate the modularity $\mathcal{Q}(\sigma_{\text{single}})$ if all r cliques form a community as expected. This is most easily calculated by taking the form provided in Eq. (2.12). The total number of links within a community is $n_c(n_c-1)/2$, so $e_c = n_c(n_c - 1)$. The degree of each node in a clique is $n_c - 1$, and the node connecting to other cliques has two additional links (that is, one incoming and one outgoing, remember we count links twice) and so has degree $n_c + 1$. The total degree K_c for each community is then $K_c = n_c(n_c - 1) + 2$. The total number of links $m = rK_c/2$ is then the sum of the degrees divided by two. Hence, we obtain

$$\begin{aligned}
\mathcal{Q}(\sigma_{\text{single}}) &= \frac{1}{2m}\sum_c [e_c - \langle e_c\rangle_{\text{conf}}] \\
&= \frac{r}{2m}\left[n_c(n_c-1) - \frac{K_c^2}{2m}\right] \\
&= \frac{r}{2m}\left[n_c(n_c-1) - \frac{K_c}{r}\right] \\
&= \frac{1}{2m}[rn_c(n_c-1) - K_c] = 1 - \frac{r}{m} - \frac{1}{r},
\end{aligned} \tag{3.1}$$

where $\langle e_c\rangle_{\text{conf}}$ denotes the expected number of edges under the configuration null-model. Now let us calculate the modularity $\mathcal{Q}(\sigma_{\text{merged}})$ if the cliques are merged two by two. So, if we originally had r communities, we will now have $r/2$ communities consisting of two adjacent cliques (assuming r is even). The number of internal edges is then $e'_c = 2e_c+2$ twice the number of edges in a single clique plus the link between the two cliques. The total degree $K'_c = 2K_c$ in a community is simply twice the total degree of a single clique. The total number of edges m of course remains unchanged. We hence arrive at

$$\begin{aligned}
\mathcal{Q}(\sigma_{\text{merged}}) &= \frac{1}{2m}\sum_c \left[e'_c - \langle e'_c\rangle_{\text{conf}}\right] \\
&= \frac{1}{2m}\frac{r}{2}\left[2n_c(n_c-1)+2-\frac{(2K_c)^2}{2m}\right] \\
&= \frac{1}{2m}\left[rn_c(n_c-1)+r-2K_c\right] = 1-\frac{r}{2m}-\frac{2}{r}.
\end{aligned} \tag{3.2}$$

The difference between the two hence becomes

$$\Delta\mathcal{Q} = \mathcal{Q}(\sigma_{\text{single}}) - \mathcal{Q}(\sigma_{\text{merged}}) = \frac{1}{2m}(K_c - r) = \frac{1}{r} - \frac{1}{K_c}$$

and if $\Delta\mathcal{Q} < 0$, or

$$K_c < r \tag{3.3}$$

then $\mathcal{Q}(\sigma_{\text{merged}})$ is the larger of the two, and hence modularity would prefer the partition of merged cliques over the partition of single cliques. For example, if we have cliques of size $n_c = 5$ and there would be more then $K_c = n_c(n_c-1)+2 = 22$ cliques, they should be merged according to modularity. For every clique size there is such a critical number of cliques above which they should be merged according to modularity. Hence, depending on the size of the graph, even the most indisputably clear communities are merged with modularity.

Alternatively, we can simply investigate the optimal number of communities for such a ring of cliques. Given that we want to minimize the number of outside links, there should be only one. Since we want to maximize the number of internal edges each community should contain the same number of links $\frac{m}{q} - 1$ for in total q communities, for which the sum of degrees equals $2\frac{m}{q}$. Hence, for q communities, the objective function value will be

$$\begin{aligned}
\mathcal{Q}(\sigma_q) &= \frac{1}{2m}q\left[2\left(\frac{m}{q}-1\right)-\frac{(2m/q)^2}{2m}\right] \\
&= \frac{1}{2m}\left[2(m-q)-2\frac{m}{q}\right] \\
&= 1-\frac{q}{m}-\frac{1}{q}.
\end{aligned} \tag{3.4}$$

We maximize with respect to q, treating the variable as continuous, and obtain

$$\frac{\partial \mathcal{Q}(\sigma_q)}{\partial q} = -\frac{1}{m} + \frac{1}{q^2}$$

so that the optimal number of communities is $q^* = \sqrt{m}$. This principle forms the basis for saying that modularity exhibits a resolution limit, and has a natural scale of $\sqrt{m}$.

This resolution limit can alternatively be interpreted as a lower bound on the community size. If we take the limit in Eq. (3.3), and consider that $K_c \approx n_c^2$ and $r = \frac{2m}{K_c}$, then any community should be at least

$$n_c \geq (2m)^{\frac{1}{4}}. \tag{3.5}$$

If a community would be smaller it would be merged by modularity, because even cliques are merged at that scale. So, this establishes the smallest community size at which communities are still "visible" to modularity.

The same idea can be generalized to communities of density p_c. Let us assume there are two subgraphs of density p_c and of equal size n_c, and that they are only linked by a single link. The total contribution when the two subgraphs are kept separate is then

$$\mathcal{Q}(\sigma_{\text{seperate}}) = \frac{1}{2m}2\left(pn_c^2 - \frac{(p_c n_c^2 + 1)^2}{2m}\right) + \mathcal{Q}_{\text{rest}}$$

while when merging them it is

$$\mathcal{Q}(\sigma_{\text{merged}}) = \frac{1}{2m}\left(2p_c n_c^2 + 2 - \frac{(2p_c n_c^2 + 2)^2}{2m}\right) + \mathcal{Q}_{\text{rest}}$$

so that the difference is

$$\Delta\mathcal{Q} = \frac{1}{2m}2\left(1 - \frac{1}{2m}(p_c n_c^2 + 1)^2\right).$$

Hence, the communities should be kept separate as long as $\Delta\mathcal{Q} < 0$, so that

$$n_c > \left(\frac{2m}{p_c^2}\right)^{\frac{1}{4}}, \tag{3.6}$$

which equals the previous bound in Eq. (3.5) when $p = 1$ of course. This shows that the resolution limit becomes more endemic if the communities are less dense than cliques.

Fig. 3.2 Ring of rings

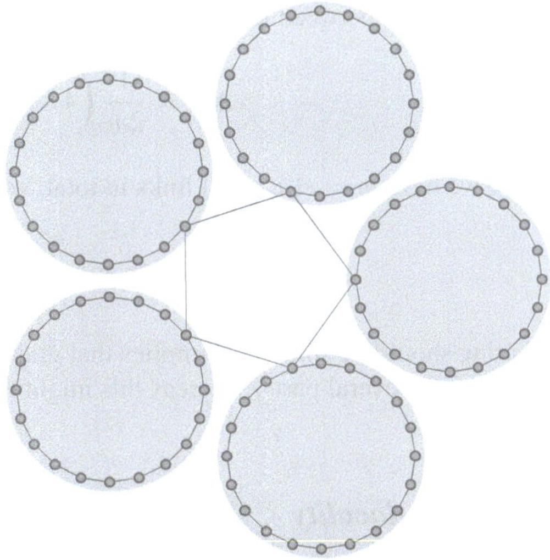

Field of View

Another problem that is related to the resolution limit is that modularity is myopic to a certain extent [4]. The graph exemplifying this problem is the rings of rings, instead of the ring of cliques, displayed in Fig. 3.2. In this case there is a central ring which connects the different rings. Especially if these rings are directed, each ring will look like a community to a random walker. After all, the probability of remaining within a single ring is substantial, with a very low probability of moving from one ring to another ring. Notice that this argument invokes a different conception of a community, namely that it is a subgraph which "traps" a random walker for a substantial amount of time, similar to Infomap [15] and the derivation by Delvenne et al. [4].

Assume we have r rings of n_c nodes, where all the rings are connected in one big ring. Then each node has degree $k_i = 2$ except for the outside node, which has $k_i = 4$. We will detail the modularity contribution for a single ring. When keeping the ring as a whole, we arrive at

$$\mathcal{Q}(\sigma_{\text{single}}) = \frac{1}{2m}\left(2n_c - \frac{(2n_c + 2)^2}{2m}\right) + \mathcal{Q}_{\text{rest}}$$

while when splitting the ring we have

$$\mathcal{Q}(\sigma_{\text{separate}}) = \frac{1}{2m}\left(4\left(\frac{n_c}{2} - 1\right) - \frac{(2\frac{n_c}{2})^2}{2m} - \frac{(2\frac{n_c}{2} + 2)^2}{2m}\right) + \mathcal{Q}_{\text{rest}}.$$

The difference between the two then amounts to

$$\Delta \mathcal{Q} = \frac{1}{2m}\left(4 - \frac{n_c^2 + 2n_c}{m}\right)$$

Since we have $m = r(n_c + 1)$ links in total, we obtain that as long as

$$4r < n_c \frac{n_c + 2}{n_c + 1}$$

this ring should be split. This implies that graphs that have long cycles or paths may be split into several parts, whereas this might not be appropriate.

3.1.2 Non-locality

Whenever a node is added, this might have effects on the other side of the network, i.e. it has a certain ripple effect [3]. More in particular, let us suppose there is a node with degree $k_i = 1$, so that it has only one neighbour (and no self-loop). Suppose the potential community to which it is linked has total degree K_c and e_c edges. Then the difference in modularity for putting the node in its neighbours community is

$$\Delta \mathcal{Q} = \frac{1}{2m}\left(2 - \frac{K_c}{m}\right) > 0,$$

so that the node should always be joined to its neighbour. A fortiori, a similar statement holds for node with any degree. Suppose some node i is the only node in a community, with degree k_i, and e_{ic} edges from i to community c. Then putting i in community c yields a benefit of

$$\Delta \mathcal{Q} = \frac{1}{2m}\left(2e_{ic} - \frac{k_i K_c}{m}\right).$$

Since $\sum_c \left[2e_{ic} - \frac{k_i K_c}{m}\right] = 0$, the difference $\Delta \mathcal{Q}$ cannot be negative for all c. So, there is at least one community to where we can move the node. Hence, no community ever consists of a single node.

This might have consequences for the rest of the partition, especially in cases where the partition is only slightly preferable to another partition. For example, suppose there are two communities which are linked strongly enough to remain together. Then if we add a single node and link it to a single node in one of the communities, they might suddenly be split. Moreover, if the other community was linked to another community they might suddenly be merged. Hence, the introduction of a single additional node might have consequences reaching beyond the local neighbourhood of the additional node, illustrated in Fig. 3.3.

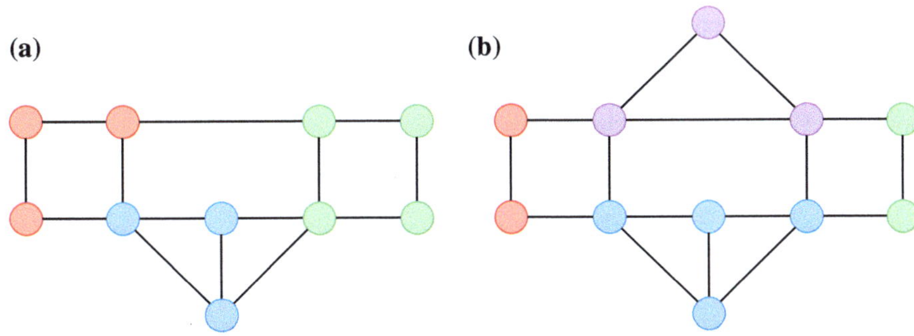

Fig. 3.3 **a** Original network. **b** Node added locality effect

3.1.3 Spuriously High Modularity

Some networks are generally believed to not contain any communities, such as random graphs, cycles and trees. Nonetheless modularity will find clusters, and modularity can be relatively high. Since modularity was in first instance a measure of the quality of the partition, it was believed that high values indicate a strongly modular structure in the network (hence the name). Furthermore, it is normalized so that for the strongest possible modular structure (the ring of cliques) modularity approaches 1. Hence, it might be expected that whenever modularity is high ($\sim$1) it is a sign that the network indeed has a significant community structure. For example, [10] suggests that values of modularity are usually somewhere ranging from 0.3 to 0.7 for networks that have some community structure.

However, modularity can reach arbitrarily high values, especially on sparse graphs [2, 9]. It can be regarded as a method that focuses on bottlenecks (few outgoing) links, while the actual density is less important [2], since this is normalized by the random null model. Hence, the value of modularity itself does not say that much, and should be interpreted with caution. We will see an illustration of how this has been wrongly applied in Chap. 6.

Let us start with the sparsest possible connected graph, a tree. A tree is minimally connected, so that if you remove any link, it will be disconnected. Furthermore, it contains no cycle, and if you add any link (without adding a node) it will contain a cycle. This implies there is a single unique path connecting any two vertices in a tree. Finally, a tree of n nodes always has $m = n - 1$ edges, which can be easily proven by induction.

Let us assume we have a tree of n nodes. We want to partition the tree along a reasonable line so as to obtain a lower bound on the modularity. Let us consider the node v which splits the network in the most equal connected components if deleted (there might be more, but let us simply choose one). Let us call that the root node of the tree. In other words, the node v which minimizes $\sum_c n_c^2$ where n_c is the size of the connected component after deletion of that node v. Let us assume this node has degree k so that there are k components. Now consider the partition that consists of these connected components and the single node v. We know that each connected subgraph of a tree is also a tree, so that each subgraph contains $e_c = 2(n_c - 1)$ edges. Counting the total degree, this is $K_c = 2m_c + 1$ because it is only connected to the root node. The total modularity is then

$$\mathcal{Q} = \frac{1}{2m}\sum_c \left[e_c - \frac{K_c^2}{2m} \right].$$

Since we only cut k links, we know that $\sum_c e_c = 2(n - 1 - k)$. Furthermore, let us assume that $n_c \approx \frac{n}{k}$ (which is a good approximation by the fact that v minimizes $\sum_c n_c^2$). We then obtain that $K_c = 2(n_c - 1) + 1 \approx 2\frac{n}{k}$ for the k components, while for the root node we obtain $K_c = k$ of course. Working out we obtain, after approximating $m \approx n$,

$$\begin{aligned} \mathcal{Q} &= \frac{1}{2m}\left[2(n - 1 - k) - \frac{1}{2m}\left(4k\frac{n^2}{k^2} - k^2 \right) \right] \\ &= 1 - \frac{k}{n} - \frac{1}{k} + \frac{k^2}{4n^2} \end{aligned} \tag{3.7}$$

so that for $n \to \infty$, we obtain that $\mathcal{Q} \to 1 - \frac{1}{k}$, assuming k remains constant. In fact, this modularity can be increased still by splitting the communities further, i.e. by recursively following the same procedure as long as it improves modularity [9]. But this changes little for the asymptotic analysis. Amazingly, this is also the modularity for the most modular network, namely the ring of cliques as mentioned in Eq. (3.1). So, according to modularity the ring of cliques and a tree are both about equally modular (for large n). Hence, even without a clear community structure the modularity is very high.

These calculations were made for a specific class of graphs, namely trees. Preferably, one would like to say that the community partition detected is significant. So, it should have a higher modularity than expected for a random graph. This is however not trivial to calculate. Using formalisms from statistical mechanics one is able to find approximate answers [12, 13], but we will not go into details here. Given a degree distribution with expected degree $\langle k \rangle$ and the average square-root degree $\langle \sqrt{k} \rangle$, the expected modularity is

$$\mathcal{Q} = 0.97\frac{\langle \sqrt{k} \rangle}{\langle k \rangle}.$$

This approximation is only good for relatively dense graphs. Whenever $\langle k \rangle$ becomes small, a random graph will contain relatively few cycles, and becomes more and more tree-like, so that the previous results apply. For ER graphs, cycles of all lengths appear simultaneously around $\langle k \rangle = 1$, and for $\langle k \rangle < 1$ it will be almost tree like.

Summarizing, upon finding some community structure, if somebody wants to comment on how "modular" the network is, the value of modularity should be compared to what can be expected in random graphs. Especially for sparse raphs modularity can become quite high, thereby making it difficult to estimate how significant a certain community structure is. For more dense graphs the modularity becomes lower, since in general it will contain more links between subgraphs.

3.2 Resolution Limit in Other Models

Most of the other models somehow try to circumvent the problem of the resolution-limit. Although not all of them were introduced specifically to deal with this issue, how they are affected by the resolution limit has been extensively analysed. Although the problem of the resolution-limit is intuitively clear—it "hides" small communities—it is not entirely clear what the opposite means. We will first discuss how the other models are affected by resolution limit like issues. We will show that most will still show similar issues. In the next section we will make more explicit what is the core of the resolution limit.

3.2.1 RB Model

Although originally not introduced in order to circumvent the problem of the resolution limit [11], the resolution parameter γ_{RB} in the Reichardt and Bornholdt model, Eq. (2.7), allows to detect communities at different scales. We repeat the same analysis as for modularity, with $p_{ij} = \frac{k_i k_j}{2m}$ but with the resolution parameter included. The objective function value for single cliques is then

$$\mathcal{H}_{\text{RB}}(\sigma_{\text{single}}) = -(rn_c(n_c - 1) - \gamma_{\text{RB}} K_c)$$

with $K_c = n_c(n_c - 1) + 2$ as before. When merging cliques we obtain

$$\mathcal{H}_{\text{RB}}(\sigma_{\text{merge}}) = -(rn_c(n_c - 1) + r - \gamma_{\text{RB}} 2K_c)$$

and so if the difference $\Delta\mathcal{H} = -\gamma_{\text{RB}} K_c + r > 0$, or

$$\gamma_{\text{RB}} K_c < r \tag{3.8}$$

it is better to merge the cliques than to separate them. The introduction of the resolution parameter γ_{RB} then changes the resolution limit as calculated for modularity. In order to keep the cliques separate the resolution parameter should be increased, while for lower resolution parameters the cliques are merged more readily. Optimizing the number of communities similar as before leads to an optimal number of communities of $q^* = \sqrt{\gamma_{\text{RB}} m}$.

The corresponding community size limit is then

$$n_c \geq \left(\frac{2m}{\gamma_{\mathrm{RB}}}\right)^{\frac{1}{4}}, \tag{3.9}$$

which can be seen from Eq. (3.8) by using $K_c \approx n_c^2$ and $r = \frac{2m}{K_c}$. Similarly for merging communities of density p_c we arrive at

$$n_c \geq \left(\frac{2m}{\gamma_{\mathrm{RB}} p_c^2}\right)^{\frac{1}{4}}. \tag{3.10}$$

Again, the introduction of the resolution parameter γ_{RB} allows to shift this lower bound on the community size, so that for a higher resolution parameter this decreases the lowest possible community size, while increasing this for lower resolution parameters.

So, introducing a resolution parameter does not circumvent completely the resolution limit, but it does allow to detect smaller or larger communities, depending on what one needs.

It is also possible to choose other null models, and a common null model is the ER null model using

$$\begin{aligned} p_{ij} = p &= \frac{m}{\binom{n}{2}} \\ &= \frac{r(n_c(n_c-1)/2+1)}{rn_c(rn_c-1)/2}. \end{aligned}$$

In this case the cost for having cliques as single communities and merging cliques is

$$\begin{aligned} \mathcal{H}_{\mathrm{RB}}(\sigma_{\mathrm{single}}) &= -(rn_c(n_c-1) - \gamma_{\mathrm{RB}} prn_c^2), \\ \mathcal{H}_{\mathrm{RB}}(\sigma_{\mathrm{merge}}) &= -(rn_c(n_c-1) + r - \gamma_{\mathrm{RB}} p\frac{r}{2}(2n_c)^2). \end{aligned}$$

The difference is then $\Delta\mathcal{H} = -\gamma_{\mathrm{RB}} rpn_c^2 + r > 0$, equivalent to $\gamma_{\mathrm{RB}} p < 1/n_c^2$ so that if

$$\gamma_{\mathrm{RB}}\,(n_c(n_c-1)+2) = \gamma_{\mathrm{RB}} K_c < r - \frac{1}{n_c},$$

the cliques should be merged. The resolution limit for the ER null model is thus the same up to a correction term of $\frac{1}{n_c^2}$. The corresponding limit on community size is

$$n_c \geq \left(\frac{1}{\gamma_{\mathrm{RB}} p}\right)^{\frac{1}{2}}. \tag{3.11}$$

More general, one can wonder if there is any null model p_{ij} such that it evades the resolution limit completely. Due to the constraint that $\sum_{ij} p_{ij} = 2m$ this is impossible however since p_{ij} scales with m. So, there will always exist a value of the resolution parameter γ_{RB} such that for some combination of q and n_c the cliques will be merged. So, even though the resolution parameter helps to look at different scales, it cannot evade the resolution limit, regardless of the null model [7].

Upper Resolution Limit

However, something interesting happens with the introduction of the resolution parameter γ_{RB}. Whereas the traditional resolution limit signifies a lower bound on the community size—communities smaller than that limit will not be detected—there are also some non-trivial upper bounds on the community size—communities larger than this limit will not be detected. This problem does not present itself in modularity (where the upper bound is trivial), and is only present in the RB model when $\gamma_{\text{RB}} > 1$. This is due to the fact that the null model then outweighs the empirical networks. This was first observed by Krings and Blondel [6] although slightly different formulated.

The problem in the traditional resolution limit is that of merging communities, but we might analyse similar limits when splitting communities. Again starting from the most modular community, a clique, we analyse when it will be split. When splitting a clique, it is best to split it into single nodes (it is better or equal to the modularity when splitting a clique in multiple parts). Let us analyse a complete clique which is completely separate from the rest of the network. Keeping the clique intact as a single community yields a cost of

$$\mathcal{H}_{\text{RB}}(\sigma_{\text{single}}) = -\left(n_c(n_c - 1) - \gamma_{\text{RB}}\frac{n_c^2(n_2 - 1)^2}{2m}\right) + \mathcal{H}_{\text{rest}}$$

while splitting it yields

$$\mathcal{H}_{\text{RB}}(\sigma_{\text{split}}) = \gamma_{\text{RB}}\frac{n_c(n_c - 1)^2}{2m} + \mathcal{H}_{\text{rest}}$$

so that we obtain that the difference is

$$\Delta\mathcal{H}_{\text{RB}} = \mathcal{H}_{\text{RB}}(\sigma_{\text{single}}) - \mathcal{H}_{\text{RB}}(\sigma_{\text{split}}) = -n_c(n_c - 1) + \frac{\gamma_{\text{RB}}}{2m}n_c^3(n_c - 1).$$

Since we want the clique not to be split, we ask that $\Delta\mathcal{H} < 0$, so that it is preferable to keep a single community. In that case, we obtain an upper bound on the community size of

$$n_c \leq \sqrt{\frac{2m}{\gamma_{\text{RB}}}}. \tag{3.12}$$

Combining the earlier result of the resolution limit [7] and this upper bound [6], we obtain the fundamental community size inequalities

$$\left(\frac{2m}{\gamma_{\text{RB}}}\right)^{\frac{1}{4}} \leq n_c \leq \left(\frac{2m}{\gamma_{\text{RB}}}\right)^{\frac{1}{2}}. \tag{3.13}$$

Notice that if $\gamma_{\text{RB}} \leq 1$ that then the upper bound indeed becomes trivial, since by definition the number of edges m is larger than the number of edges in the clique of size $n_c^2/2$, and so that by definition $n_c \leq \sqrt{2m}$. Whenever $\gamma_{\text{RB}} \geq 2m$ the lower bound becomes trivial, since $n_c \geq 1$ by definition. Of course, when $\gamma_{\text{RB}} \geq 2m$ then also $n_c \leq 1$ by the upper bound, so that necessarily $n_c = 1$. In fact, when $\gamma_{\text{RB}} \geq 2m$ the inequalities conflict, and the statement is no longer valid. So, only for $1 < \gamma_{\text{RB}} < 2m$ both bounds are non-trivial and valid, and reduce the size of "visible" communities to the ranges specified.

Instead of separating cliques, let us investigate when it is beneficial to split communities of a lower density p. Suppose we have a subgraph of density p which is difficult to split. That is, let us suppose that any partition in two creates subgraphs that have (about) the same density p, and also a density of about p between the two subgraphs. Let us first consider what would be the contribution of this subgraph to modularity. The number of internal edges is then pn_c^2, which is of course equal to the total degree. We then arrive at a cost of

$$\mathcal{H}_{\text{RB}}(\sigma_{\text{single}}) = -\left(pn_c^2 - \frac{(pn_c^2)^2}{2m}\right) + \mathcal{H}_{\text{rest}}.$$

When splitting the graph in two, the number of internal edges is then $p\left(\frac{n_c}{2}\right)^2$, while the total degree is $\frac{pn_c^2}{2}$. Hence, the contribution for splitting is

$$\mathcal{H}_{\text{RB}}(\sigma_{\text{split}}) = -2\left(p\left(\frac{n_c}{2}\right)^2 - \frac{\left(\frac{pn_c}{2}\right)^2}{2m}\right) + \mathcal{H}_{\text{rest}}.$$

Examining the difference we arrive at

$$\Delta\mathcal{H}_{\text{RB}} = -\frac{pn_c^2}{2}\left(1 - \gamma_{\text{RB}}\frac{pn_c^2}{2m}\right),$$

and so we obtain the upper bound on the community size of

$$n_c < \sqrt{\frac{2m}{\gamma_{\text{RB}}\,p}}. \tag{3.14}$$

Notice that this coincides with the original upper bound when $p = 1$. Combining the lower and upper bound, we obtain

$$\left(\frac{2m}{\gamma_{\text{RB}}}\right)^{\frac{1}{4}} \leq \sqrt{p}n_c \leq \left(\frac{2m}{\gamma_{\text{RB}}}\right)^{\frac{1}{2}}. \tag{3.15}$$

Notice again, that these bounds are only non-trivial and valid for $1 \leq \gamma_{\text{RB}} \leq 2m$.

Surprisingly, when considering the ER null model no such upper bound exists. If we work out the case of splitting a complete clique, we arrive at the inequality that

$$n_c(n_c - 1)(1 - \gamma_{\text{RB}} p) > 0. \tag{3.16}$$

This is always the case when $\gamma_{\text{RB}} < 1/p$ and the clique will never be split, while if $\gamma_{\text{RB}} > 1/p$ the clique will always be split. Hence, there is no particular community size n_c for which it will be split or not.

This points to an interesting difference between the lower bound and upper bound resolution limit. Whereas the lower bound holds regardless of any null model, the upper bound holds only for certain null models.

3.2.2 AFG Model

The AFG model was introduced specifically to overcome to some extent the resolution limit inherent in modularity [1]. Similar to the RB model, varying the parameter γ_{AFG} allows one to obtain different views of the community structure, although the two methods are not equivalent (except trivially when $\gamma_{\text{AFG}} = 0$ and $\gamma_{\text{RB}} = 1$, in which case both reduce to modularity). The question is to what extent this method is able to overcome the resolution limit.

Let us start again with the traditional ring of cliques. The AFG model can alternatively be written as

$$\mathcal{H}_{\text{AFG}} = -\sum_c \left[e_c + \gamma_{\text{AFG}} n_c - \frac{(K_c + \gamma_{\text{AFG}} n_c)^2}{2m + \gamma_{\text{AFG}} n} \right].$$

For the ring of cliques we have $e_c = n_c(n_c - 1)$ while $K_c = e_c + 2$. The cost for having each clique as a community then is

$$\mathcal{H}_{\text{AFG}} = -r \left(e_c + \gamma_{\text{AFG}} n_c - \frac{(K_c + \gamma_{\text{AFG}} n_c)^2}{2m + \gamma_{\text{AFG}} n} \right)$$

while for merging the cliques we have

$$\mathcal{H}_{\text{AFG}} = -\frac{r}{2} \left(2(e_c + 1 + \gamma_{\text{AFG}} n_c) - \frac{4(K_c + \gamma_{\text{AFG}} n_c)^2}{2m + \gamma_{\text{AFG}} n} \right)$$

and the difference comes down to

$$\Delta\mathcal{H} = r \left(1 - \frac{(K_c + \gamma_{\text{AFG}} n_c)^2}{2m + \gamma_{\text{AFG}} n} \right).$$

Since $2m = rK_C$ and $n = rn_c$ we obtain that the cliques will be merged ($\Delta\mathcal{H} > 0$) when

$$K_c + \gamma_{\text{AFG}} n_c < r \tag{3.17}$$

which is the original limit in Eq. (3.3) up to a correction of $\gamma_{\text{AFG}} n_c$ of the usual resolution limit of $\gamma_{\text{AFG}} n_c$. While this correction was multiplicative using γ_{RB}, it is additive using γ_{AFG}. The lower bound on the community size is implicitly given by

$$(\gamma_{\text{AFG}} + n_c) n_c^3 > 2m \tag{3.18}$$

which for $n_c \gg \gamma_{\text{AFG}}$ becomes equivalent to the original inequality of $n_c \geq (2m)^{\frac{1}{4}}$. On the other hand if $n_c \approx \gamma_{\text{AFG}}$ this amounts to $n_c \geq m^{\frac{1}{4}}$. So we generally expect the smallest community size to scale as $m^{\frac{1}{4}}$.

Upper Resolution Limit

Let us now analyse whether the AFG model also exhibits an upper bound on the community size. We know that for $\gamma_{\text{AFG}} = 0$ no such bound exists, but for the RB model such a bound exists for $\gamma_{\text{RB}} > 1$, so perhaps for $\gamma_{\text{AFG}} > 0$ this is also the case.

Let us again start from a clique completely separate from the rest of the network. This then amounts to

$$\mathcal{H}_{\text{AFG}}(\sigma_{\text{single}}) = n_c(n_c - 1) + \gamma_{\text{AFG}} n_c - \frac{n_c(n_c - 1) + \gamma_{\text{AFG}} n_c)^2}{2m + \gamma_{\text{AFG}} n} + \mathcal{H}_{\text{rest}}$$

while splitting into single nodes yields

$$\mathcal{H}_{\text{AFG}}(\sigma_{\text{split}}) = n_c \left(\gamma_{\text{AFG}} - \frac{(n_c - 1 + \gamma_{\text{AFG}})^2}{2m + \gamma_{\text{AFG}} n} \right) + \mathcal{H}_{\text{rest}}.$$

So the difference becomes

$$\Delta\mathcal{H}_{\text{AFG}} = n_c(n_c - 1) \left(1 - \frac{(n_c - 1 + \gamma_{\text{AFG}})^2}{2m + \gamma_{\text{AFG}} n} \right),$$

which gives the upper bound on the community size of

$$n_c \leq \sqrt{2m + \gamma_{\text{AFG}} n} - \gamma_{\text{AFG}} + 1. \tag{3.19}$$

This bound is only non-trivial for $0 < \gamma_{\text{AFG}} < \frac{-n + \sqrt{8m + n^2}}{2}$, since by definition $1 < n_c < \sqrt{2m}$. Hence, this restricts the community sizes also to some specific range depending on γ_{AFG}.

3.2.3 CPM and RN

The method of CPM [16] and RN [14] differ from the other two models in the sense that they do not depend on any null model. For non-weighted graphs it can be easily seen that their definitions are equal, and so we will only state the results for CPM here, which is somewhat more elegant in its presentation. The corresponding inequalities for the RN model can be derived using $\gamma_{\text{CPM}} = \frac{\gamma_{\text{RN}}}{1+\gamma_{\text{RN}}}$. Finally, the LP method corresponds to $\gamma_{\text{CPM}} = 0$.

We again start out by looking at when cliques are merged in the ring of cliques network. We again have r cliques of size n_c connected to each other with only a single link. The cost for keeping them separate is then

$$\mathcal{H}_{\text{CPM}}(\sigma_{\text{single}}) = -r(n_c(n_c - 1) - \gamma_{\text{CPM}} n_c^2)$$

while when merging them it is

$$\begin{aligned}\mathcal{H}_{\text{CPM}}(\sigma_{\text{merge}}) &= -\frac{r}{2}(2n_c(n_c - 1) + 2 - \gamma_{\text{CPM}} 4n_c^2) \\ &= -r(n_c(n_c - 1) + 1 - \gamma_{\text{CPM}} 2n_c^2).\end{aligned}$$

The difference is

$$\Delta\mathcal{H}_{\text{CPM}} = r(1 - \gamma_{\text{CPM}} n_c^2),$$

so that the cliques will be merged when $\gamma_{\text{CPM}} < 1/n_c^2$. As one can see, this no longer depends on the total size of the network in terms of m or n, but only on the "local" variable n_c. However, the lower bound on the community size still exists, which is

$$n_c > \sqrt{\frac{1}{\gamma_{\text{CPM}}}}. \tag{3.20}$$

The question remains what exactly the resolution limit entails and whether the independence of this inequality on m and n suffices to say a method does not suffer from the resolution limit. We will come back to this question in the next section.

We can also look at when two subgraphs of size n_c and density p with only a single link in between them will be merged. Doing so yields a cost of

$$\mathcal{H}_{\text{CPM}}(\sigma_{\text{single}}) = 2(pn_c^2 - \gamma_{\text{CPM}} n_c^2) + \mathcal{H}_{\text{rest}}$$

for keeping them separate. Merging them gives a cost of

$$\mathcal{H}_{\text{CPM}}(\sigma_{\text{merged}}) = (2pn_c^2 + 1 - \gamma_{\text{CPM}}(2n_c)^2) + \mathcal{H}_{\text{rest}}$$

so that the difference yields a lower bound on the community size of

$$n_c > \sqrt{\frac{1}{\gamma_{\text{CPM}}}}. \tag{3.21}$$

Surprisingly this bound remains unchanged for different densities. This implies that the community sizes only depend on the resolution parameter γ_{CPM} and not on the graph analysed.

More general, suppose there are two communities of sizes n_1 and n_2, with a density of $2pn_1n_2$ links in between, so that the density is p, with each e_1 and e_2 links within the communities, and let us see when they should be merged. Keeping them separate yields a cost of

$$\mathcal{H}_{\text{CPM}}(\sigma_{\text{single}}) = e_1 + e_2 - \gamma_{\text{CPM}}(n_1^2 + n_2^2) + \mathcal{H}_{\text{rest}}$$

while, merging them gives a cost of

$$\mathcal{H}_{\text{CPM}}(\sigma_{\text{merged}}) = (e_1 + e_2 + pn_1n_2 - \gamma_{\text{CPM}}(n_1 + n_2)^2) + \mathcal{H}_{\text{rest}}$$

so that the two should be merged whenever

$$\gamma_{\text{CPM}} < p. \tag{3.22}$$

This simply states that two communities should always be merged whenever the density of the links between them exceeds the resolution parameter. Vice-versa, the communities should be separated when the density is below this threshold.

This provides a quite clear interpretation and definition of a community. If we consider only changes to the partition that consist of moving (a set of) nodes from one community to another, merging communities and splitting communities—in other words, a local minimum of most algorithms considered here—a community should then have

1. a uniform internal density of at least γ_{CPM}; and
2. a uniform external density of at most γ_{CPM} to each other community.

With a uniform internal density, we mean that for all partitions of the community, there will be at least a density of γ_{CPM} link between them. In other words, we should not be able to split a community in two, so that for all proper subsets $S \subset C$ where C is the community set of some community, we should have

$$\frac{e_{S,C'}}{|S|(|C| - |S|)} > \gamma_{\text{CPM}} \tag{3.23}$$

where $e_{S,C'}$ is the number of edges between S and $C' = S \backslash C$ its complement in C. With a uniform external density, we mean that we cannot merge a subset of one community to another community. Suppose $S \subset C$ and we consider another community D, then it should hold that

$$e_{S,C'} - \gamma|S|(|C| - |S|) > e_{S,D} - \gamma|S|(|D| + |S|). \tag{3.24}$$

If not, we could put set S in community D.

This already provides some intuition as to the upper resolution limit. Since each community should have an internal density of at least γ_{CPM}, a clique will not be quickly split. Performing the same calculations as before for splitting a clique into separate nodes yields the inequality

$$n_c(n_c - 1)(1 - \gamma_{\text{CPM}}) < 0 \tag{3.25}$$

which is always satisfied for $\gamma_{\text{CPM}} < 1$ and never satisfied for $\gamma_{\text{CPM}} > 1$. Indeed, this corresponds nicely with the definition of a community just provided. This implies there is no upper resolution limit.

However, a particular problem for CPM is that communities of different densities are difficult to detect simultaneously. Suppose there is a rather sparse community with uniform density p, and two cliques separated by a density of p as well. Then whenever $\gamma_{\text{CPM}} > p$ it should merge the two cliques, while if $\gamma_{\text{CPM}} < p$ it should split the sparse community. Hence, it is impossible to find a single γ_{CPM} such that all communities are detected correctly. On the other hand, when a community has a density p and the cliques are separated with a density of p, it is also a valid question whether they actually constitute good communities. One possible solution is to have somehow different resolution parameters γ_{ij} per link (i, j).

Summarizing, the bounds for CPM (and by extension RN) do not depend on the actual graph, and there is no upper bound on the community size. This indeed suggests that these methods are less troubled by the resolution limit. In that sense they might be preferable to the other methods, although this might change depending on the needs.

3.3 Scale Invariance

In the previous section we have detailed quite specifically what the different bounds of the different methods are, detailed in Table 3.1. However, it remains somewhat unclear to what extent these methods suffer from the resolution limit. More specifically, when does a method *not* suffer from the resolution limit? So the concept of *resolution limit free* requires a more precise definition. We will develop such a definition in this section, and see it has a natural connection with the scale of the graph. Methods that not suffer from the resolution limit are said to be *scale invariant*. Moreover, we have seen that some models suffer from a lesser extent to the resolution limit. A natural question is what models do not suffer from the resolution limit? Phrased somewhat differently, what weights a_{ij} and b_{ij} can we choose in the general framework in order to be scale invariant?

Table 3.1 Resolution limits of different methods

Method	Lower resolution limit ($n_c \geq \cdots$)	Upper resolution limit ($n_c \leq \cdots$)
Modularity	$(2m)^{\frac{1}{4}}$	–
RB with configuration null-model	$\left(\frac{2m}{\gamma_{\text{RB}}}\right)^{\frac{1}{4}}$	$\left(\frac{2m}{\gamma_{\text{RB}}}\right)^{\frac{1}{2}}$
RB with ER null-model	$\left(\frac{1}{p\gamma_{\text{RB}}}\right)^{\frac{1}{2}}$	–
AFG[a]	$(\gamma_{\text{AFG}} + n_c)n_c^3 > 2m$	$\sqrt{2m + \gamma_{\text{AFG}} n} - \gamma_{\text{AFG}} + 1$
RN	$\left(\frac{1+\gamma_{\text{RN}}}{\gamma_{\text{RN}}}\right)^{\frac{1}{2}}$	–
CPM	$\left(\frac{1}{\gamma_{\text{RB}}}\right)^{\frac{1}{2}}$	–
LP	n	–

[a]Lower resolution limit only given implicitly

3.3.1 Relaxing the Null Models

One of the reasons that modularity and the RB and AFG models suffer from a resolution limit is their dependence on a null model. In the canonical derivation of our framework, it was demanded that $\sum_{ij} p_{ij} = 2m$. This makes sense, since from the point of view of a random null model we would like to have as many edges in the random graph as in the original graph. However, from the view point of simply specifying some of the weights a_{ij} and b_{ij} we are not constrained a-priori.

So, let us relax that constraint for a moment. Consider for example that we take away the dependence on the number of links in the configuration null model, so that we take $p_{ij} = k_i k_j$. Of course, this has little sense in terms of a null model, but it simply corresponds to a weight of $b_{ij} = \gamma_{\text{RB}} k_i k_j$. In other words, the cost of having a missing link in a community is proportional to how many links both nodes have. This is no less arbitrary then simply choosing a $b_{ij} = |N(i) \cup N(j)|$ the number of neighbours in common for example.

Let us briefly see how such a method using $p_{ij} = k_i k_j$ would perform on the ring of cliques. Keeping the cliques separate yields

$$\mathcal{H}_{\text{RB}}(\sigma_{\text{single}}) = -r(e_c - \gamma_{\text{RB}} K_c^2)$$

while merging them yields

$$\mathcal{H}_{\text{RB}}(\sigma_{\text{merge}}) = -\frac{r}{2}(2e_c + 2 - \gamma_{\text{RB}}(2K_c)^2) = -r(e_c + 1 - \gamma_{\text{RB}} 2K_c^2)$$

with the difference

$$\Delta\mathcal{H}_{\text{RB}} = 1 - \gamma_{\text{RB}} K_c^2 > 0$$

so that the cliques are merged whenever

$$\gamma_{\mathrm{RB}} < K_c^{-2}. \tag{3.26}$$

This also does not depend on the graph under consideration, so this model could also be said not to suffer from the resolution limit. But is this really what we mean by not suffering from the resolution limit?

Not all problems have disappeared. Suppose we take the subgraph consisting of only two of these cliques. We analyse when the method would merge the two cliques in this subgraph. The only difference is that in this subgraph $K_c' = K_c - 1$ because there is only a single link connecting the two subgraphs. The two cliques will then be merged when $\gamma_{\mathrm{RB}} < K_c'^2 = (K_c - 1)^2$. Even though neither inequality depends on any global variables, a problem remains. Combining the above two inequalities, we obtain that whenever

$$K_c^{-2} < \gamma_{\mathrm{RB}} < (K_c - 1)^{-2}, \tag{3.27}$$

the method will separate the cliques in the larger graph, yet merge them in the subgraph. So even though the inequality for merging cliques in Eq. (3.26) does not depend on any global variable, some problems of scale remain.

Considering again modularity (or the RB model) this is similar to what happens in the resolution limit. Since the merging of cliques will depend on the size of the graph as a whole, indeed cliques will be merged in some large graph, while in the subgraph they will not be merged. Taking the same subgraph of two cliques, and taking $K_c' = K_c$ as before, the cliques will be merged in the large graph and split in the subgraph when

$$\frac{r}{K_c - 1} < \gamma_{\mathrm{RB}} < \frac{r}{K_c}. \tag{3.28}$$

3.3.2 Defining Scale Invariance

The above discussion motivates the following idea for a scale invariant method. The general idea is that when looking at any induced subgraph of the original graph, the resulting partitioning should not change. In order to introduce this definition, let $\mathcal{H}$ be any objective function (which we want to minimize), we then call a partition $\mathcal{C}$ for a graph $\mathcal{H}$-optimal whenever $\mathcal{H}(\mathcal{C}) \leq \mathcal{H}(\mathcal{C}')$ for any other partition $\mathcal{C}'$. We can then define *scale invariance* as follows.

Definition 3.1 Let $\mathcal{C} = \{C_1, C_2, \dots, C_q\}$ be an $\mathcal{H}$-optimal partition of a graph G. Then the objective function $\mathcal{H}$ is called *scale invariant* if for each subgraph H induced by $\mathcal{D} \subset \mathcal{C}$, the partition $\mathcal{D}$ is also $\mathcal{H}$-optimal.

Intuitively this means the following. If we take any subgraph induced by the optimal partition, that same partition should be optimal on that subgraph. Since the subgraph is *induced* by the optimal partition, it can only consist of complete communities; we can't cut across any communities. This idea is illustrated in Fig. 3.4.

Notice that this definition only "works" in one direction. That is, if a method is scale invariant, we know that for all subgraphs induced by the optimal partition, the partition remains optimal on those subgraphs. However, the inverse is obviously not true. Suppose we are given some graphs $G_1, \ldots, G_q$ and some optimal partitions on them $\mathcal{C}_1, \ldots, \mathcal{C}_q$. Then an arbitrary graph G which has all graphs $G_i \subset G$ as a subgraph, does not necessarily have the same optimal partition composed of $\mathcal{C}_1, \ldots, \mathcal{C}_q$.

But for the ring of cliques this one direction is all we need. Observe that indeed when a method merges cliques in the ring of cliques network depending on the size of the network it is not scale invariant. To see this, it is slightly easier to consider the chain of cliques—the ring of cliques with one link between two cliques cut out. For some number of cliques r above some threshold $r > r^*$ it will merge cliques, while for $r' < r^*$ it will keep them separate. The graph with r' cliques is a subgraph of G_r, or $G'_r \subset G_r$ which is induced by its partition of merged cliques $\mathcal{C}$. Indeed the partition $\mathcal{D}$ inducing graph G'_r is no longer optimal, since they are no longer merged in graph G'_r. Finally, notice that if a method is scale invariant, it will never start merging cliques depending on the number of cliques r. So, this definition indeed accurately captures the core of the resolution limit altogether. This idea is demonstrated in Fig. 3.5. In short, the same partition should then remain optimal for that induced subgraph.

If the objective function has another property, we can state something interesting, namely that to some extent we can go to the other "direction". That is, we may exchange parts of optimal partitions. So if we have an optimal partition of the complete graph G, and we find another optimal partition on some (community) induced subgraph H, we may exchange it.

Definition 3.2 An objective function $\mathcal{H}$ for a partition $\mathcal{C} = \{C_1, \ldots, C_q\}$ is called additive whenever $\mathcal{H}(\mathcal{C}) = \sum_i \mathcal{H}(C_i)$, where $\mathcal{H}(C_i)$ is the objective function defined on the subgraph H induced by C_i.

Notice that CPM and the RN model are both additive objective functions, but that modularity, the RB and the AFG model are not. Although the first is easy to see, the latter is perhaps less clear. The essential notion here is that $\mathcal{H}(C_i)$ is defined on the subgraph H induced by C_i, so that it may not depend on anything outside the subgraph H. The latter models already depend on some global parameters, but even the local dependence on the degree k_i renders these method not additive.

Now the interesting result is that if we have an $\mathcal{H}$-optimal partition $\mathcal{C}$ for an additive scale invariant objective function $\mathcal{H}$, we can replace subpartitions of $\mathcal{C}$ by other optimal subpartitions, as already stated informally.

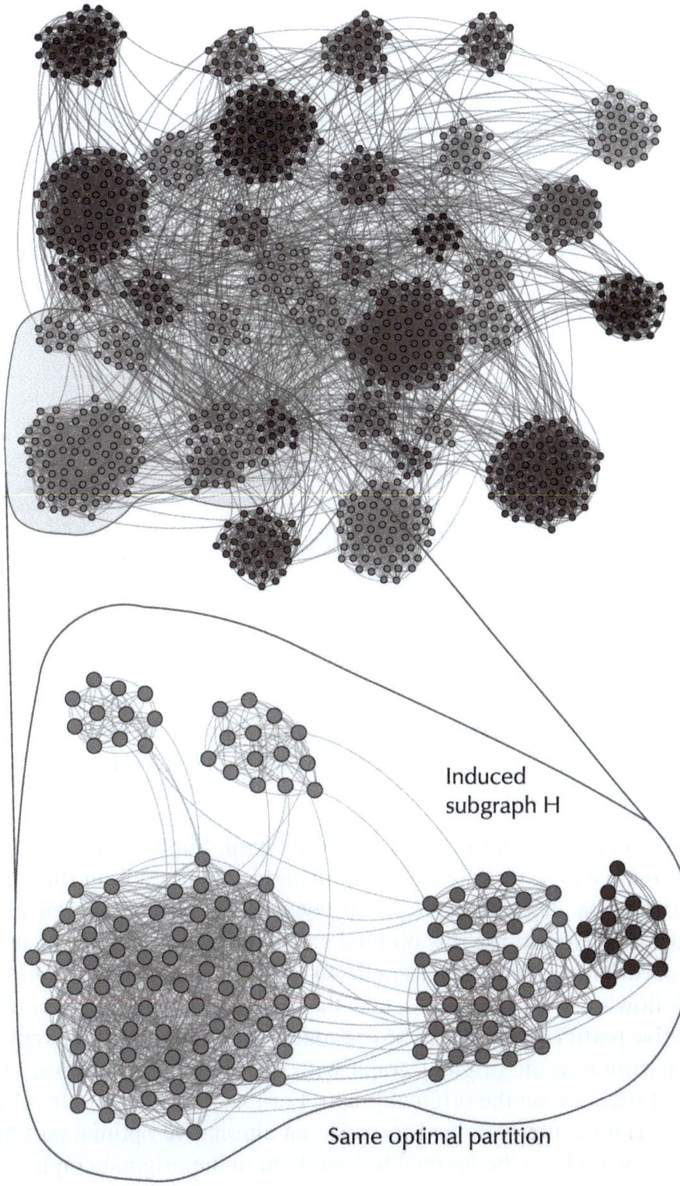

Fig. 3.4 Scale invariance illustration

Theorem 3.3 *Given an additive scale invariant objective function $\mathcal{H}$, let $\mathcal{C}$ be an $\mathcal{H}$-optimal partition of a graph G and let $H \subset G$ be the induced subgraph by $\mathcal{D} \subset \mathcal{C}$. If $\mathcal{D}'$ is an alternative optimal partition of H then $\mathcal{C}' = \mathcal{C} \setminus \mathcal{D} \cup \mathcal{D}'$ is also $\mathcal{H}$-optimal.*

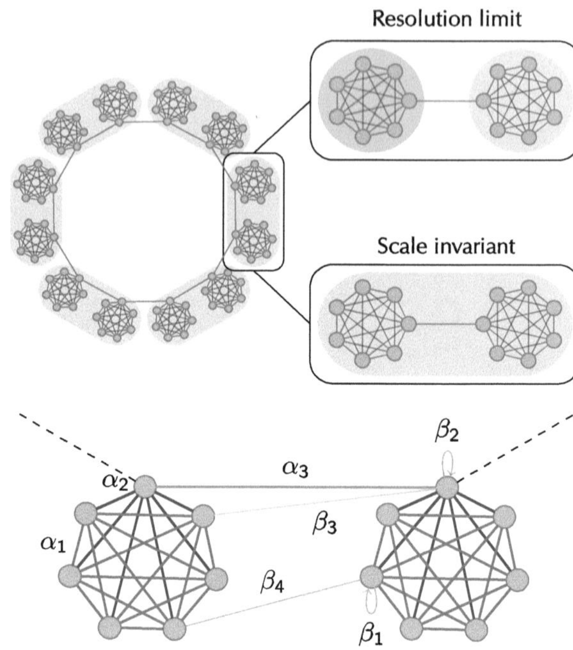

Fig. 3.5 Scale invariance in ring of cliques

Proof Define $\mathcal{C}'$ and $\mathcal{D}'$ as in the theorem. By additivity, $\mathcal{H}(\mathcal{C}') = \mathcal{H}(\mathcal{C} \setminus \mathcal{D}) + \mathcal{H}(\mathcal{D}')$, and by optimality $\mathcal{H}(\mathcal{D}') \leq \mathcal{H}(\mathcal{D})$. Since also $\mathcal{H}(\mathcal{C}) = \mathcal{H}(\mathcal{C} \setminus \mathcal{D}) + \mathcal{H}(\mathcal{D})$ we obtain $\mathcal{H}(\mathcal{C}') \leq \mathcal{H}(\mathcal{C})$, so $\mathcal{C}'$ is also optimal. □

The idea behind this proof is simply the following. Suppose we have an optimal partition $\mathcal{C}$. Then suppose we take a community induced subgraph and have a different optimal partition on that subgraph. Then because of the property of an additive objective function, we can use this optimal partition on the subgraph to replace that part of the partition on the original graph. In terms of the example in Fig. 3.4, this means the following. Suppose that we take the subgraph H as indicated in the figure. If an alternative partition would also be optimal on that subgraph, then replacing that part of the partition in the original graph with the alternative partition would also be an optimal partition on the original graph. For example, if the four communities in the bottom right could be joined to create an alternative optimal partition in the subgraph H, it would also be optimal to join them in the original graph.

Although this might seem to contradict the NP-hardness of community detection methods, this is not the case. It states that when there are two optimal partitions, any combination of those partitions are optimal, so in a certain sense, they are spanning a space of optimal partitions. It does not say whether such a partition can be easily found. Also, there might be two optimal partitions that cannot be obtained by recombining them, because all communities partly overlap with each other. For example, suppose that one optimal $\mathcal{C}$ partition is to divide the set of nodes in C_1 and C_2, while another optimal partition is $\mathcal{C}' = \{C_1', C_2'\}$, where $C_i \cap C_j' \neq \emptyset$. Then both partitions give rise to different induced subgraphs, hence the one cannot be used to replace parts of the other.

We can prove that CPM is *scale invariant* in this sense, just like the RN model and the LP model. The RB model is not scale invariant according to our definition, regardless of the null model [7], and hence modularity is not scale invariant. Furthermore, as we have seen, also when using $p_{ij} = k_i k_j$ the model is not scale invariant. Finally, the AFG model is not scale invariant either.

For scale invariant methods, the results should be unchanged on subgraphs. Hence, we could try to run an algorithm recursively on subgraphs. We could for example consider the following improvement for CPM. First we cut the network at each recursive call, until the density of the subgraph exceeds γ_{CPM}. Then, we recombine the subgraphs, and loop over nodes/communities to find improvements until we can no longer increase greedily, and return to the previous recursive function call. These calls should be easily parallelized, making community detection in even larger graphs or in an on-line setting possible.

Since the CPM model is also related to the RB model using the ER null model, it is tempting to conclude it is also scale invariant. Indeed, this might be said to be the case, if we choose p independently of the graph, i.e. not define it as $p = m/n(n-1)$, and simply choose it as some value $p \in \mathbb{R}$. However, we then obviously retrieve the CPM model. This shows that scale invariant methods are strongly constrained, and there is only a fine line between resolution-limit and scale invariant methods.

These results follow from the more general theorem we will now prove. For this, we first introduce the notion of local weights. Again, building on the idea of subgraphs, we define local weights as weights that do not change when looking to subgraphs.

Definition 3.4 Let G be a graph, and let a_{ij} and b_{ij} as in Eq. (2.5) be the associated weights. Let H be a subgraph of G with associated weights a'_{ij} and b'_{ij}. Then the weights are called *local* if $a_{ij} = \lambda a'_{ij}$ and $b_{ij} = \lambda b'_{ij}$, where $\lambda = \lambda(H) > 0$ can depend on the subgraph H.

Notice that this multiplicative scaling with λ leaves unchanged the optimum of the objective function $\mathcal{H}$. Clearly then, the RN and CPM model have local weights, while the RB and AFG model do not. This definition says that local weights should be independent of the graph G in a certain sense. In fact, it is quite a strong requirement, as it should even hold for a single link (ij) in the subgraph where only i and j are included. That means it can not depend on any other link *but* the very link itself. Since for missing links, there is (usually) no associated weight or anything, it can only be constant. There are some exceptions, such as multi-partite networks, or networks embedded in geographical space [5, 8], where some meaningful non-constant local weights can be provided. Hence, the RN model and the CPM model are one of the few sensible options available for having local variables. We can now prove the more general statement that methods using local weights are scale invariant.

Theorem 3.5 *The objective function $\mathcal{H}$ as defined in* Eq. (2.5) *is* scale invariant *if it has* local *weights.*

Proof Let $\mathcal{C}$ be the optimal partition for G with community assignments c_i, $\mathcal{D} \subset \mathcal{C}$ a subset of this partition, and H the subgraph induced by $\mathcal{D}$ with h nodes. Furthermore, we denote by d_i the community indices of $\mathcal{D}$, such that $d_i = c_i$ for $1 \leq i \leq h$ and by A' the adjacency matrix of H, so that $A_{ij} = A'_{ij}$ for $1 \leq i \leq h$. Assume $\mathcal{D}$ is not optimal for H, and that $\mathcal{D}^*$ is optimal, so that $\mathcal{H}(\mathcal{D}) > \mathcal{H}(\mathcal{D}^*)$. Then define c^* by setting $c_i^* = d_i^*$ for $1 \leq i \leq h$ and $c_i^* = c_i$ for $h < i \leq n$. Then because the result is unchanged for the nodes $h < i \leq n$, we have that

$$\Delta\mathcal{H} = \mathcal{H}(\mathcal{C}) - \mathcal{H}(\mathcal{C}^*) = \frac{1}{\lambda}(\mathcal{H}(\mathcal{D}) - \mathcal{H}(\mathcal{D}^*)) > 0$$

where the last step follows from the locality of the weights a_{ij} and b_{ij}. This inequality contradicts the optimality of $\mathcal{C}$. Hence, for all induced subgraphs H, the partition $\mathcal{D}$ is optimal, and the objective function $\mathcal{H}$ is *scale invariant*. □

The converse is unfortunately not true. Consider a graph G with some weights a_{ij} and b_{ij}. Then pick a subgraph H induced by some subpartition $\mathcal{D}$, and define the weights $a'_{ij} = a_{ij}$ and $b'_{ij} = b_{ij}$ except for one particular edge (kl), for which we set $a'_{kl} = a_{kl} + \epsilon$. Then for some $\epsilon > 0$, the original subpartition will remain optimal in H, while the weights are not local. Since the small change of the weight is *only* made when considering the graph H, all other subpartitions will always remain optimal. Of course, such a definition of the weight is rather odd, so in practice we will never use it.

Even though the converse is not true, we can say a bit more. The weights can be a bit different indeed, but there is not that much room for these differences. We demonstrate this on the ring of cliques. The weights can depend only on the graph, so if G and G' are two isomorphic graphs (i.e. they are the same up to a relabelling of the nodes), then $a_{ij}(G) = a_{i'j'}(G')$, where i and i' are two isomorphic nodes. Hence, only a number of weights can be different from each other in the ring network, as illustrated in Fig. 3.5. All nodes within a clique are isomorphic, except the node that connects to other cliques. So, all the edges among those $n_c - 1$ nodes are similar, and will have the same weight α_1. All edges from these $n_c - 1$ nodes to the "outside" node will have the same weight α_2. Finally, the edge connecting two cliques is denoted by α_3. The missing self-loop for the special outside node is denoted by β_2 while the missing self-loop for the other nodes in the cliques is denoted by β_1. Finally, there is (1) a missing link between the outside node and a normal node denoted by β_3; and (2) a missing link between two normal nodes, denoted by β_4. These weights are illustrated in Fig. 3.5.

Let us now analyse when a method is not scale invariant. Then, the cliques must be merged in some (large) graph, while for the subgraph consisting of these two merged cliques, they should be separated by the method. Or conversely, they should be separated in some (large) graph, but merged in the subgraph. We can write the $\mathcal{H}(\sigma_{\text{separate}})$ for all r cliques being separate as

$$\mathcal{H}(\sigma_{\text{separate}}) = -r(\alpha_1(n_c-1)(n_c-2) + 2\alpha_2(n_c-1) \\ - (n_c-1)\beta_1 - \beta_2)$$

and $\mathcal{H}(\sigma_{\text{merged}})$ for merging all two consecutive cliques as

$$\mathcal{H}(\sigma_{\text{merged}}) = -\frac{r}{2}2(\alpha_1(n_c-1)(n_c-2) + 2\alpha_2(n_c-1) \\ - (n_c-1)\beta_1 - \beta_2 + \alpha_3 - \beta_3(n_c-1) - \beta_4(n_c-1)^2)$$

Furthermore, for the induced subgraph H consisting of two consecutive cliques, we can write $\mathcal{H}'_s$ for separating the two cliques and $\mathcal{H}'_m$ for merging them, similarly as before, where α' and β' are the weights for the subgraph H. Then the method is not scale invariant if it would merge the two cliques at a higher level (i.e. when $\mathcal{H}_m < \mathcal{H}_s$) yet would not merge them at smaller scale (i.e. when $\mathcal{H}'_s < \mathcal{H}'_m$), or vice versa. Working out this condition for $\mathcal{H}_m < \mathcal{H}_s$ (and similarly for $\mathcal{H}_m > \mathcal{H}_s$) gives us

$$\alpha_3 > (n_c-1)(\beta_4(n_c-1) + \beta_3),$$

while for $\mathcal{H}'_s < \mathcal{H}'_m$ (and similarly for $\mathcal{H}'_s > \mathcal{H}'_m$) we obtain

$$\alpha'_3 < (n_c-1)(\beta'_4(n_c-1) + \beta'_3).$$

Combining these two inequalities for both cases we obtain

$$\alpha'_3(\beta_4(n_c-1) + \beta_3) < \alpha_3(\beta'_4(n_c-1) + \beta'_3), \tag{3.29}$$

$$\alpha'_3(\beta_4(n_c-1) + \beta_3) > \alpha_3(\beta'_4(n_c-1) + \beta'_3). \tag{3.30}$$

where either Eq. (3.29) or (3.30) should hold. Hence, only if the left hand side equals the right hand side, it does not constitute a counter example. Working out this equality, there are two possibilities. Either the weights should be local, or the following equality should hold

$$n_c - 1 = \frac{\alpha_3\beta'_3 - \alpha'_3\beta_3}{\alpha'_3\beta_4 - \alpha_3\beta'_4}. \tag{3.31}$$

Obviously, this again constitutes some very particular case of non-local weights. We can repeat this same procedure for other subpartitions, and for other graphs, thereby forcing the weights to be of a very particular kind. This thus leaves little room for having any sensible non-local definition such that the method is scale invariant.

This means scale invariant community detection has only a quite limited scope. In fact, CPM seems to be the simplest non-trivial sensible formulation of any general scale invariant method, although there is some leeway for special graphs (i.e. having some node properties, such as multi-partite graphs). This is not to say that methods with non-local weights (e.g. modularity, AFG, number of triangles, shortest path, betweenness) should never be used for community detection at all, they are just never scale invariant.

References

1. Arenas A, Fernandez A, Gomez S (2007) Analysis of the structure of complex networks at different resolution levels. New J Phy 10(5):23. doi:10.1088/1367-2630/10/5/053039. arXiv:physics/0703218
2. Bagrow J (2012) Communities and bottlenecks: trees and treelike networks have high modularity. Phys. Rev. E 85(6):9. doi:10.1103/PhysRevE.85.066118. arXiv:1201.0745
3. Brandes U, Delling D, Gaertler M, Gorke R, Hoefer M et al (2008) On modularity clustering. IEEE Trans Knowl Data Eng 20(2):172–188. doi:10.1109/TKDE.2007.190689
4. Delvenne JC, Yaliraki SN, Barahona M (2010) Stability of graph communities across time scales. Proc Natl Acad Sci USA 107(29):12755–12760. doi:10.1073/pnas.0903215107
5. Expert P, Evans TS, Blondel VD, Lambiotte R (2011) Uncovering space-independent communities in spatial networks. Proc Natl Acad Sci USA 108(19):7663–7668. doi:10.1073/pnas.1018962108. arXiv:1012.3409v1
6. Krings G, Blondel VD (2011) An upper bound on community size in scalable community detection. arXiv:1103.5569
7. Kumpula JM, Saramäki J, Kaski K, Kertész J (2007) Limited resolution in complex network community detection with Potts model approach. Eur Phys J B 56(1):41–45. doi:10.1140/epjb/e2007-00088-4
8. Lambiotte R, Blondel VD, De Kerchove C, Huens E, Prieur C et al (2008) Geographical dispersal of mobile communication networks. Physica A 387(21):5317–5325. doi:10.1016/j.physa.2008.05.014
9. Montgolfier FD, Soto M, Viennot L (2011) Asymptotic modularity of some graph classes. Lecture notes in computer science, vol 7074, pp 435–444. doi:10.1007/978-3-642-25591-5_45
10. Newman M, Girvan M (2004) Finding and evaluating community structure in networks. Phys Rev E 69(2):026113. doi:10.1103/PhysRevE.69.026113
11. Reichardt J, Bornholdt S (2006a) Statistical mechanics of community detection. Phys Rev E 74(1):016110. doi:10.1103/PhysRevE.74.016110
12. Reichardt J, Bornholdt S (2006) When are networks truly modular? Physica D 224(1–2):20–26. doi:10.1016/j.physd.2006.09.009
13. Reichardt J, Bornholdt S (2007) Partitioning and modularity of graphs with arbitrary degree distribution. Phys Rev E 76(1):015102. doi:10.1103/PhysRevE.76.015102
14. Ronhovde P, Nussinov Z (2010) Local resolution-limit-free Potts model for community detection. Phys Rev E 81(4):046114. doi:10.1103/PhysRevE.81.046114. arXiv:0803.2548v4
15. Rosvall M, Bergstrom CT (2008) Maps of random walks on complex networks reveal community structure. Proc Natl Acad Sci USA 105(4):1118–1123. doi:10.1073/pnas.0706851105
16. Traag VA, Van Dooren P, Nesterov Y (2011) Narrow scope for resolution-limit-free community detection. Phys Rev E 84(1):016114. doi:10.1103/PhysRevE.84.016114. arXiv:1104.3083

Chapter 4
Finding Significant Resolutions

In this section we will focus on how to determine correct resolution parameters for CPM. Although for the other methods it might also be relevant to determine resolution parameters, they do have some natural resolution parameter, although it might not be the "best" one. For the RB model this is $\gamma_{\text{RB}} = 1$ and for the AFG model $\gamma_{\text{AFG}} = 0$, in which case both of course reduce to modularity. Although different resolution parameters might be chosen—and in the light of the resolution limits perhaps even should be chosen—there at least exists some natural resolution parameter around which to try other resolutions. For CPM this is not the case, since we simply choose some constant γ_{CPM}, and so we should need some additional effort in determining when a certain resolution parameter "works well". Since we will only use the CPM method here, we will use $\gamma_{\text{CPM}} = \gamma$ to avoid cluttering the notation.

4.1 Scanning Resolution Parameter

Although we do not have any a-priori idea about a specific resolution parameter, there do exist some simple bounds of course. If we set $\gamma = 0$ the only trivial solution is all nodes in a single community. This can also be seen from the condition of merging communities in Eq. (3.22). Assuming the graph to be connected, we know that there are always at least two communities that have at least one edge, and so that the density p between the two communities is at least $p > 0 = \gamma$, and so by Eq. (3.22) the two communities should be merged. Since this condition remains true until all communities are merged, we are left with a single large community. The objective function value at this point is then $\mathcal{H} = -\sum_{ij} A_{ij} = -2m$.

On the other hand, if $\gamma = 1$ all nodes will be in a single community. Applying the merging condition in Eq. (3.22) again, starting from each node in its own community, it is immediately clear that two nodes should never be merged, since any density p between two communities is never larger than 1, and so $p \leq 1 = \gamma$. The objective function value at this point is then $\mathcal{H} = -\sum_i (A_{ii} - \gamma) = n$ assuming there are no self loops. So, we know that $\gamma \in [0, 1]$ for unweighed graphs.

V. Traag, *Algorithms and Dynamical Models for Communities and Reputation in Social Networks*, Springer Theses, DOI: 10.1007/978-3-319-06391-1_4,

However, choosing a specific γ is not straightforward. Commonly, it is assumed that "good" partitions are somehow "stable". For example, one could perturb slightly the network to see if the partition remains the same [6]. Another possibility is stability with respect to various stochastic runs [9]. If the algorithm returns partitions that are very different, the partition is not very stable, and we might hence expect that the partition is not very good. As before, we could measure the similarity between two partitions using either NMI or VI. Since the VI is not normalized, it is a bit more sensitive to any fluctuations in the partition, and so preferable for this task. So, we might run the algorithm multiple times and see how stable the results are using the VI.

Another approach would be to look at the stability of the partition with respect to the resolution parameter γ [4]. So, if a partition remains optimal over some relatively large range $[\gamma_1, \gamma_2]$, it should indicate it is a relatively good partition. We know that for a certain specific γ communities should have an inner density higher than γ and the density between any two communities is lower than γ. If the same partition remains stable over the range of $[\gamma_1, \gamma_2]$ then we know that the communities have inner density γ_2 (for $\gamma > \gamma_2$ CPM splits communities) and are separated by a density of γ_1 (for $\gamma < \gamma_1$ CPM merges communities). Hence, the larger this range over which the partition remains stable, the more clear-cut the community structure. Moreover, it is the ratio γ_2/γ_1 between the two parameters that counts, not the absolute difference $\gamma_2 - \gamma_1$. After all, if a communities have a density of $\gamma_2 = 0.80$ and are separated by $\gamma_1 = 0.75$ this is not quite the same as having a density of $\gamma_2 = 0.1$ and separated by $\gamma_1 = 0.05$. Hence, we will usually plot in logarithmic scale.

However, if we need to scan the whole range of $\gamma \in [0, 1]$ to some granularity, and also rerun the algorithm multiple times for all values we want to check, this becomes quite computationally intensive. Fortunately, we need not check all values $\gamma \in [0, 1]$. This is readily clear because there are finitely many partitions, while there are infinitely many values of γ. But there is an even stronger property, namely that the optimal solutions remain optimal for some range [5, 7]. More precisely, if σ is an optimal solution for γ_1 and γ_2, then σ is also an optimal solution for all $\gamma \in [\gamma_1, \gamma_2]$.

Theorem 4.1 *Let $\mathcal{H}(\gamma, \sigma)$ be the CPM objective function. Then if*

$$\sigma^* = \arg\max_\sigma \mathcal{H}(\gamma_1, \sigma) = \arg\max_\sigma \mathcal{H}(\gamma_2, \sigma)$$

then $\sigma^* = \arg\max_\sigma \mathcal{H}(\gamma, \sigma)$ for $\gamma_1 \leq \gamma \leq \gamma_2$.

Proof This is a result of the linearity of $\mathcal{H}(\gamma, \sigma)$ in γ. To see this, suppose that σ^* is optimal in γ_1 and γ_2. Let $\gamma = \lambda\gamma_1 + (1 - \lambda)\gamma_2$ with $0 \leq \lambda \leq 1$, then by linearity of $\mathcal{H}(\gamma, \sigma)$ in γ we have

$$\mathcal{H}(\gamma, \sigma^*) = \lambda\mathcal{H}(\gamma_1, \sigma^*) + (1 - \lambda)\mathcal{H}(\gamma_2, \sigma^*).$$

Since $\mathcal{H}(\gamma_1, \sigma^*) \leq \mathcal{H}(\gamma_1, \sigma)$ and $\mathcal{H}(\gamma_2, \sigma^*) \leq \mathcal{H}(\gamma_2, \sigma)$ for any σ,

$$\mathcal{H}(\gamma, \sigma^*) \leq \lambda\mathcal{H}(\gamma_1, \sigma) + (1 - \lambda)\mathcal{H}(\gamma_2, \sigma) = \mathcal{H}(\gamma, \sigma),$$

and so $\mathcal{H}(\gamma, \sigma^*) \leq \mathcal{H}(\gamma, \sigma)$ and σ^* is optimal for $\gamma \in [\gamma_1, \gamma_2]$.

That $\mathcal{H}(\gamma, \sigma)$ is linear in γ can be seen from the definition. Slightly rewritten we obtain

$$\begin{aligned}\mathcal{H}(\gamma, \sigma) &= -\sum_{ij}(A_{ij} - \gamma)\delta(\sigma_i, \sigma_j) \\ &= -(E - \gamma N)\end{aligned} \tag{4.1}$$

where $E := \sum_c e_c$ the total of internal edges and $N := \sum_c n_c^2$ is the sum of the squared community sizes, and it is immediately clear this is linear in γ. □

Moreover, it turns out that N is also monotonically decreasing with γ. This makes sense, since with increasing γ, more and more weight is put on N, so to minimize $\mathcal{H}$ smaller values of N are needed. This corresponds also to finding smaller communities with increasing γ. Notice that for $\gamma = 0$ we have $N = n^2$, while for $\gamma = 1$ we have that $N = n$.

Theorem 4.2 *Let* $\sigma_z = \arg\max_\sigma \mathcal{H}(\gamma_z, \sigma)$, $z = 1, 2$. *Furthermore, let* $N_z = \sum_c n_c^2(\sigma_z)$ *where* $n_c(\sigma_z)$ *denote the community sizes of the partition* σ_z. *If* $\gamma_1 \leq \gamma_2$ *then* $N_1 \geq N_2$.

Proof Let there be two different optimal partitions σ_1^* and σ_2^* for $\gamma_1 < \gamma_2$, with costs

$$\begin{aligned}\mathcal{H}(\gamma_1, \sigma_1^*) &= -E_1 + \gamma_1 N_1, \\ \mathcal{H}(\gamma_2, \sigma_2^*) &= -E_2 + \gamma_2 N_2.\end{aligned}$$

Then since both partitions are optimal for the corresponding resolution parameters we obtain

$$\begin{aligned}-E_1 + \gamma_1 N_1 &\leq -E_2 + \gamma_1 N_2, \\ -E_2 + \gamma_2 N_2 &\leq -E_1 + \gamma_2 N_1.\end{aligned}$$

Summing both inequalities, we obtain

$$-(E_1 + E_2) + \gamma_1 N_1 + \gamma_2 N_2 \leq -(E_1 + E_2) + \gamma_1 N_2 + \gamma_2 N_1$$

and so $\gamma_1(N_1 - N_2) \leq \gamma_2(N_1 - N_2)$, and since $\gamma_1 < \gamma_2$ we obtain that $N_1 \geq N_2$. □

Notice that if both partitions are optimal for both resolution parameters, then necessarily $N_1 = N_2$, and so also $E_1 = E_2$. Hence, any two equally good partitions, must have the same number of internal edges and squared community sizes.

So, denoting by $N(\gamma)$ the sum of squared community sizes corresponding to the optimal partition for some γ, we obtain that $N(\gamma)$ is a monotonically decreasing function. Because $N(\gamma_1) = N(\gamma_2)$ if a partition is optimal for both γ_1 and γ_2, this is a stepwise monotonically decreasing function. The minimum $\min_\sigma \mathcal{H}(\gamma, \sigma)$ is hence a piecewise linear monotonically increasing function.

Hence, we only need to find those points at which $N(\gamma)$ changes, which can be done reasonably effectively using bisectioning on γ. Let us assume we start on some interval $[\gamma_1, \gamma_2]$. If $N(\gamma_1) \neq N(\gamma_2)$, we know that $N(\gamma) = N(\gamma_1)$ for some γ between γ_1 and γ_2. So, we can recursively split the interval to check for this γ, up to some $|\gamma_2 - \gamma_1| > \delta$, or on logarithmic scale.

Algorithm 7 Recursive bisectioning of the resolution parameter

```
function ResBisect(γ1, γ2, map N)
   if |N(γ1) − N(γ2)| > ε  |γ1 − γ2| > δ then
      γmid ← (γ1+γ2)/2
      N(γmid) ← CommDetect(γmid)
      ResBisect(γ1, γmid, N)
      ResBisect(γmid, γ2, N)
   end if
end function
function GetRes
   N ← empty map
   ResBisect(0, 1, N, orn logarithmic scale)
   return N
end function
```

In addition, if we run multiple times the community detection on the values of γ found by this algorithm, we also have some indication of the stability of the partition.

Unfortunately, many networks results are often messy, so that it still remains a challenge to choose a "correct" resolution parameter. Nonetheless, this method seems to work quite well on benchmark networks, as displayed in Fig. 4.1. These benchmark networks have $n = 10^3$ nodes and have an average degree $\langle k \rangle = 10$ with a maximum degree of $\Delta = 50$. The community sizes range between 10 and 100. The exponent of the power law distribution of both the community sizes and the degree sizes was set at $\tau = 2$. It is quite clear that N is stepwise decreasing, and $\mathcal{H}$ piecewise linear increasing. The plateaus (indicated by magenta) indeed correspond to the planted partition for the benchmark network. The resolution parameter γ^*_{CPM} we used for testing is also displayed. For $\mu = 0.1$ this parameter falls nicely in the plateau, but for $\mu = 0.5$ the parameter is slightly off. In addition, in the range of the plateau, the VI is relatively low (near 0), indicating the partition is relatively stable. Hence, using such heuristics, it seems possible to scan for "stable" plateaus of resolution values.

Even though this might point to resolution parameters γ for which the partitions are somehow "good", this does not say they are significant. Moreover, we cannot say anything about which resolution level is preferable in some way, and all partitions are (to some extent) valid partitions of the network. So, we cannot say anything about

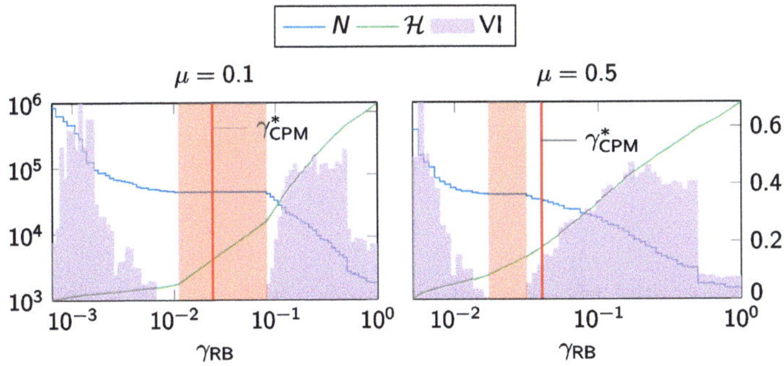

Fig. 4.1 Scanning resolution parameter

the "correct" or "true" partition, unless there is very clearly only a single resolution parameter which (almost) always returns the same partition. After some reflection, it is ironic we return to the question of what resolution returns a good partition. After all, the initial goal of modularity was in fact to state what partition is especially good.

4.2 Significance of Partition

Although modularity compares the number of edges within a community to a random graph, this does not provide any significance of a partition, since random graphs also can have quite high modularity. When thinking about the significance of a partition, modularity goes about it the wrong way around. We do *not* want to know the probability that random edges fall within the found communities, as done by modularity. Nonetheless, explicitly calculating the actual probability that a partition is as dense as detected seems to yield good results [1, 2]. Rather, we are interested in the probability that such a dense partition can be found in a random graph. Comparing the observed modularity to the expected modularity for a random graph gives some idea of the significance of a partition [8]. But preferably this should be made more specific. More in particular, let E be again the number of edges within communities for the whole partition. Then we are interested in the probability that a random graph (with the same number of edges and nodes) contains a partition with at least E edges within communities. Notice that this is quite different from the probability that a random partition contains at least E edges, which is basically what modularity does.

Unfortunately, this probability is reasonably difficult to calculate, but we can break it down in some parts. Let us focus on the question of the probability of finding a certain dense subgraph within a random graph. Once we have the probability of finding a certain dense subgraph, we should be able to apply this recursively on the remainder of the graph and partition. That is, once we know the probability to find a community C, we look at the complementary graph with nodes $V \setminus C$, and ask what the probability is to find some community in that graph. Unfortunately, we cannot provide the exact probabilities, but obtain some insightful asymptotic results. In addition, the dominant terms of the asymptotic results suggest some approximation of the probability of finding a dense partition in some random graph. This can be used to determine which resolution parameters are significant, in addition to the previously discussed results on stable plateaus.

4.2.1 Preliminaries

We are interested in estimating the probability that a certain subgraph is contained in a random graph. When speaking of a subgraph, we usually mean an induced subgraph, that is a subset of nodes, with all the edges of the that subset present in the subgraph. More specifically, a subgraph H such that $V(H) \subset V(G)$ and that $E(H) = \{(i, j) \in E(G) \mid i, j \in V(H)\}$. So, we are given a certain number of vertices n_c and edges m_c and are asked what the probability is that a random graph contains an induced subgraph of the specified order with that many edges. For some specific subgraphs, we need to take into account isomorphisms. Here we are only interested in subgraphs with some number of edges, so that we do not need to address this issue.

We write $G \in \mathcal{G}(n, p)$ for a random graph G from $\mathcal{G}(n, p)$, such that each edge has independent probability p of being included in the graph, the usual ER graphs. In this section we will use the notation $|G| = |V(G)| = n$ for the number of nodes and $\|G\| = |E(G)| = m$ for the number of edges. We write $\Pr(H \subseteq \mathcal{G}(n, p))$ for the probability that H is an induced subgraph of a $G \in \mathcal{G}(n, p)$. Notice that we will always use $H \subseteq G$ to denote the fact that H is an *induced* subgraph of G. Let $S(n_c, m_c) = \{G \mid |G| = n_c, \|G\| = m_c\}$ denote the set of all graphs with $n_c = |G|$ vertices and $m_c = \|G\|$ edges. Furthermore, we use a bit of abusive notation and write $\Pr(S(n_c, m_c) \subseteq \mathcal{G}(n, p))$ for the probability that a graph $G \in \mathcal{G}(n, p)$ contains one of the graphs in $S(n_c, m_c)$, i.e.

$$\Pr(S(n_c, m_c) \subseteq \mathcal{G}(n, p)) = \Pr\left(\bigcup_{H \in S(n_c, m_c)} H \subseteq \mathcal{G}(n, p)\right).$$

Notice that whenever $n = n_c$ we obtain the ordinary binomial probabilities

$$\Pr(S(n_c, m_c) \subseteq \mathcal{G}(n_c, p)) = \binom{\binom{n_c}{2}}{m_c} p^{m_c}(1-p)^{\binom{n_c}{2}-m_c}.$$

Let us denote by X the random variable that represents the number of occurrences of a subgraph with n_c vertices and m_c edges in a random graph. Let X_H be the indicator value that specifies whether a subgraph H of order $n_c = |H|$ in the random graph equals one of the graphs in $S(n_c, m_c)$, so that

$$X_H = \begin{cases} 1 & \text{if } \exists H \in S(n_c, m_c) \\ 0 & \text{otherwise} \end{cases}$$

which of course comes down to

$$X_H = \begin{cases} 1 & \text{if } \|H\| = m_c \text{ and } |H| = n_c \\ 0 & \text{otherwise} \end{cases}.$$

We can then write $X = \sum_H X_H$ where the sum runs over all $\binom{n}{n_c}$ possible subgraphs H. Obviously then, X is a non-negative random variable and $\Pr(X > 0) = \Pr(S(n_c, m_c) \subseteq \mathcal{G}(n, p))$. We will rely on two useful inequalities: Markov's inequality

$$\Pr(X \geq a) \leq \frac{\mathbb{E}(X)}{a},$$

which for us will be most useful in the form $\Pr(X > 0) = \Pr(X \geq 1) \leq \mathbb{E}(X)$, and Chebyshev's inequality

$$\Pr(|X - \mathbb{E}(X)| \geq \lambda) \leq \frac{\mathbb{E}(X^2) - \mathbb{E}(X)^2}{\lambda^2},$$

or in a form more useful to us

$$\Pr(X = 0) \leq \Pr\left[|X - \mathbb{E}(X)| \geq \mathbb{E}(X)\right] \leq \frac{\mathbb{E}(X^2) - \mathbb{E}(X)^2}{\mathbb{E}(X)^2}.$$

This way of estimating probabilities is known as the second moment method.

4.2.2 Subgraph Probability

We here present bounds for estimating the probability of a subgraph with n_c vertices and m_c edges occurring in a random graph. We start off with a particular easy one. But first, let us analyse $\mathbb{E}(X)$. Here it is convenient to define

$$r = \Pr(S(n_c, m_c) \subseteq \mathcal{G}(n_c, p)).$$

Theorem 4.3 *The expected number of occurrences of an induced subgraph can be written as*

$$\mathbb{E}(X) = \binom{n}{n_c} r \tag{4.2}$$

Proof By linearity of expectation, we have $\mathbb{E}(X) = \sum_H \mathbb{E}(X_H)$, and because X_H is an indicator variable $\mathbb{E}(X_H) = \Pr(X_H = 1)$. Notice that H has n_c nodes, so that $H \in \mathcal{G}(n_c, p)$, and $\Pr(X_H = 1) = r$. There are $\binom{n}{n_c}$ subgraphs of n_c nodes in a graph with n nodes, which concludes the proof. □

Using Markov's inequality, this leads to the following bounds.

Theorem 4.4 *The probability that $\mathcal{G}(n, p)$ contains an induced subgraph with n_c nodes and m_c edges can be bounded by*

$$1 - (1-r)^{\lfloor \frac{n}{n_c} \rfloor} \leq \Pr\left(S(n_c, m_c) \subseteq \mathcal{G}(n, p)\right) \leq \binom{n}{n_c} r.$$

Proof The upper bound is immediate from Theorem 4.3 and Markov's inequality. For the lower bound, consider a partition of $G \in \mathcal{G}(n, p)$ into equal parts $H_1, \ldots, H_k$ of size n_c, for which $k = \lfloor n/n_c \rfloor$. Again, for each of these parts, the probability to have m_c edges $\Pr(X_{H_i} = 0) = 1 - r$, and the probability that none of these parts have m_c edges is $(1-r)^{\lfloor n/n_c \rfloor}$, since they are independent. Hence, the probability that $\Pr(X = 0) \leq (1-r)^{\lfloor n/n_c \rfloor}$. □

In order to improve on the lower bound, we need to calculate $\mathbb{E}(X^2)$. The idea here is to calculate the expected value of the number of pairs of subgraphs that have m_c edges. If the two subgraphs are independent this average is fairly simple to calculate. In the case of overlap between the two this is more complicated. Nonetheless, we can then separate into three parts: the parts of the two subgraphs without overlap, and the part that overlaps. Working this out brings us the following lower bound.

Theorem 4.5 *The probability that $\mathcal{G}(n, p)$ contains no induced subgraph with n_c nodes and m_c edges can be bounded by*

$$\Pr(S(n_c, m_c) \nsubseteq \mathcal{G}(n, p)) \leq \frac{1}{\mathbb{E}(X)} \sum_{u \geq 1} \binom{n_c}{u} \binom{n - n_c}{n_c - u}$$
$$\sum_{m(\Delta)}^{\min(\binom{u}{2}, m_c)} \binom{M(u)}{m_c - m(\Delta)} p^{m_c - m(\Delta)} (1-p)^{M(u) - (m_c - m(\Delta))}. \tag{4.3}$$

with $M(u) = \frac{n_c(n_c - 1) - u(u-1)}{2}$.

Proof The variable X^2 can be decomposed into parts $X_H \times X_{H'}$, such that we need to investigate the probability that both H and H' have m_c edges. So, we can separate this expectancy in parts of partially overlapping subgraphs, like

$$\mathbb{E}(X^2) = \sum_u \sum_{|H \cap H'| = u} \Pr(\|H\| = \|H'\| = m_c), \tag{4.4}$$

where u represents the overlap between the different subgraphs. If H and H' are (edge) independent, so when $u < 1$, the answer is simple, and is given by $\Pr(X_H = 1)^2$. For $u \geq 1$ the answer is more involved.

So let us consider two subgraphs H and H' such that $|H \cap H'| = u \geq 1$. Let us separate this in three independent parts, the overlap $\Delta = H \cap H'$, and the remainders $A = H - \Delta$ and $B = H - \Delta$. Clearly then, $|\Delta| = u$, and $|A| = |B| = n_c - u$. The probability that $\|H\| = \|H'\| = m_c$ can then be decomposed in the probability that the sum of these independent parts sum to m. The probability that $\|H\| = m_c$ can be decomposed as

$$\Pr(\|H\| = m_c) = \sum_{m(\Delta)} \Pr(\|\Delta\| = m(\Delta))$$
$$\Pr(\|H\| = m_c \mid \|\Delta\| = m(\Delta)).$$

where $m(\Delta)$ signifies the number of edges within Δ. Similarly, we arrive at the conditional probability for both subgraphs H and H'. However, since we have conditioned exactly on the overlapping part, the two remaining parts are independent, and we can write

$$\begin{aligned}\Pr(\|H\| = \|H'\| = m_c \mid \|\Delta\| = m(\Delta))\\ = \Pr(\|H\| = m_c \mid \|\Delta\| = m(\Delta))^2.\end{aligned}$$

This probability can be calculated and yields

$$\begin{aligned}&\Pr(\|H\| = m_c \mid \|\Delta\| = m(\Delta))\\ &= \binom{M(u)}{m_c - m(\Delta)} p^{m_c - m(\Delta)} (1-p)^{M(u) - (m_c - m(\Delta))},\end{aligned}$$

where $M(u) = \frac{n_c(n_c-1) - u(u-1)}{2}$. We then obtain

$$\begin{aligned}\Pr(\|H\| = \|H'\| = m_c) = &\sum_{m(\Delta)} \Pr(\|\Delta\| = m(\Delta))\\ &\binom{M(u)}{m_c - m(\Delta)}^2 p^{2(m_c - m(\Delta))} (1-p)^{2M(u) - 2(m_c - m(\Delta))}\end{aligned}$$

which leads to

$$\binom{\binom{n_c}{2}}{m_c} p^{m_c} (1-p)^{\binom{n_c}{2} - m_c} \sum_{m(\Delta)} \binom{M(u)}{m_c - m(\Delta)} p^{m_c - m(\Delta)} (1-p)^{M(u) - (m_c - m(\Delta))},$$

where $m(\Delta)$ ranges from 0 to the minimum of m_c and the number of possible edges $\binom{u}{2}$.

Now counting the number of subgraphs that overlap in u nodes, for each choice of subgraph H, we choose u nodes in H, and $n_c - u$ nodes in the remaining $n - n_c$ nodes. In total, there are then

$$C_u = \binom{n}{n_c}\binom{n_c}{u}\binom{n-n_c}{n_c-u}$$

overlapping subgraphs with u nodes in common. Concluding, we arrive at

$$\mathbb{E}(X^2) = \sum_u C_u \Pr\left(\|H\| = \|H'\| = m_c \mid |H \cap H'| = u\right).$$

Writing this out, we see that

$$\mathbb{E}(X^2) = \mathbb{E}(X) \sum_{u \geq 1} \binom{n_c}{u}\binom{n-n_c}{n_c-u}$$
$$\sum_{m(\Delta)}^{\min(\binom{u}{2}, m_c)} \binom{M(u)}{m_c - m(\Delta)} p^{m_c - m(\Delta)} (1-p)^{M(u)-(m_c-m(\Delta))} + \mathbb{E}(X)^2,$$

where the last term $\mathbb{E}(X)^2$ comes from the non-overlapping part. Working out the Chebyshev inequality, we obtain the inequality stated in the theorem. □

4.2.3 Asymptotic Analysis

We focused on subgraphs of a fixed size n_c in the previous section. However, for asymptotic analysis, this is not interesting, as it is already clear that all fixed size subgraphs are contained in the random graph asymptotically. So let us consider subgraphs of size proportional to n, so that it is of size sn, with $0 < s < 1$. Of course, then m_c should also grow accordingly, and we consider the subgraph with a fixed density q. For the asymptotic analysis, we can afford to be a bit sloppy with this density, and consider $(sn)^2$ possible edges in the subgraph of sn nodes, so that $m_c = q(sn)^2$, and we now denote by $S(n, q)$ the subgraphs with density q instead of the actual number of edges. Using the previously calculated bounds, we can then prove the following asymptotic statement.

Theorem 4.6 *Asymptotically almost surely, no graph contains subgraphs of size sn, with $0 < s < 1$, with density $q \neq p$, and will contain subgraphs of density $q = p$ of any size,* i.e.

$$\lim_{n\to\infty} \Pr(S(sn, q) \subseteq \mathcal{G}(n, p)) = \begin{cases} 0 & \textit{if } p \neq q \\ 1 & \textit{if } p = q \end{cases} \tag{4.5}$$

Proof We will first prove the 0-statement, for which the upper bound suffices. Applying Stirling's formula to $\binom{n}{n_c}$ we obtain

$$\binom{n}{n_c} \sim \frac{\sqrt{n}}{\sqrt{2\pi n_c(n-n_c)}} \exp\left(nH\left(\frac{n_c}{n}\right)\right)$$
$$= \frac{1}{\sqrt{2\pi s(1-s)n}} \exp(nH(s)),$$

where $H(p)$ is the binary entropy

$$H(p) = -p\log p - (1-p)\log(1-p). \tag{4.6}$$

Working out yields

$$\mathbb{E}(X) \sim \frac{\exp\left[nH(s) + (sn)^2 H(q)\right]}{2\pi sn\sqrt{s(1-s)q(1-q)n}} p^{q(sn)^2}(1-p)^{(1-q)(sn)^2},$$

or

$$\mathbb{E}(X) \sim \frac{\exp\left[nH(s) - (sn)^2 D(q,p)\right]}{2\pi sn\sqrt{s(1-s)q(1-q)n}},$$

utilising the binary Kullback-Leibler divergence [3]

$$D(q,p) = q\log\frac{q}{p} + (1-q)\log\frac{1-q}{1-p}. \tag{4.7}$$

Since $D(q,p) > 0$ for $p \neq q$ we can conclude that $\mathbb{E}(X) \to 0$ for $n \to \infty$ when $p \neq 0$.

We need the second moment for the lower bound. This can be rewritten as $\sum_u \sum_{m(\Delta)} f(u, m(\Delta))$ with

$$f(u, m(\Delta)) = \frac{\binom{sn}{u}\binom{(1-s)n}{sn-u}\binom{(sn)^2-u^2}{q(sn)^2-m(\Delta)} p^{-m(\Delta)}(1-p)^{-u^2+m(\Delta)}}{\binom{n}{sn}\binom{(sn)^2}{q(sn)^2}}$$

Using the notation $u = \alpha sn$ and $q(sn)^2 - \Delta = \beta((sn)^2 - u^2)$ this becomes

$$f(\alpha,\beta) = \frac{\binom{sn}{\alpha sn}\binom{(1-s)n}{s(1-\alpha)n}\binom{(1-\alpha^2)(sn)^2}{\beta(1-\alpha^2)(sn)^2}}{\binom{n}{sn}\binom{(sn)^2}{q(sn)^2}}$$
$$p^{-(q-\beta(1-\alpha^2))(sn)^2}(1-p)^{(q-\beta(1-\alpha^2)-\alpha^2)(sn)^2}$$

Taking logarithms on Sterling's approximation, we obtain

$$\log \binom{n}{k} = \mathcal{O}\left(nH\left(\frac{k}{n}\right)\right),$$

Applying this approximation, we obtain

$$\begin{aligned}\log(f(\alpha,\beta)) = \mathcal{O}\Big[n\left(sH(\alpha) + (1-s)H\left(\frac{(1-\alpha)s}{1-s}\right) - H(s)\right)\\ + n^2s^2\left((1-\alpha^2)H(\beta) + (q-\beta(1-\alpha^2))\log\frac{1-p}{p}\right.\\ \left. - \alpha^2\log(1-p) - H(q)\right)\Big].\end{aligned}$$

Using again the binary Kullback-Leibler divergence $D(p, q)$, we can simplify this to

$$\log(f(\alpha,\beta)) = \mathcal{O}\left[nQ + n^2s^2(D(q,p) - (1-\alpha^2)D(\beta,p))\right],$$

with $Q = \left(sH(\alpha) + (1-s)H\left(\frac{(1-\alpha)s}{1-s}\right) - H(s)\right)$. The range over which α and β can vary are as follows. Since u ranged from 1 to $n_c = sn$, α ranges from 0 to 1. The range of β depends on α:

$$\beta \in \begin{cases} \left[\frac{q-\alpha^2}{1-\alpha^2}, \frac{q}{1-\alpha^2}\right] & \text{if } \alpha^2 < \text{q} \\ \left[0, \frac{q}{1-\alpha^2}\right] & \text{if } \alpha^2 \geq \text{q} \end{cases}.$$

Notice that we are interested in the case that $p = q$, so that $D(q, p) = 0$. Then $D(\beta, p) > 0$ for $\alpha < 1$ because of the range of β, and $Q < 0$ if $\alpha = 1$ because $H(s) > 0$ for $0 < s < 1$, so that $\Pr(X = 0) \to 0$ as $n \to \infty$ for $p = q$. □

This suggests that any partition (in a finite number of communities) of the random graph will asymptotically contain only communities of density approximately p. This matches some results on community detection on random graphs using CPM. Whenever $\gamma < p - \epsilon$ only a single community will be detected, while for $\gamma > p + \epsilon$ only communities consisting of only single nodes will be detected, while for $p - \epsilon \leq \gamma \leq p + \epsilon$ a transition takes place where several communities are detected with density approximately p. This transition interval shrinks with increasing n, so that $\epsilon \to 0$ for $n \to \infty$, consistent with the asymptotic analysis provided here. The only difference of course is that for γ relatively high, we start to divide into a number of communities that grows with n, so that the limit is no longer correct. Nonetheless, it explains reasonably well the transition, and is illustrated in Fig. 4.2. Here, the transition becomes more clear for larger graphs, and approaches asymptotically the limit at $\gamma = p$ indicated by the dotted line in the figure.

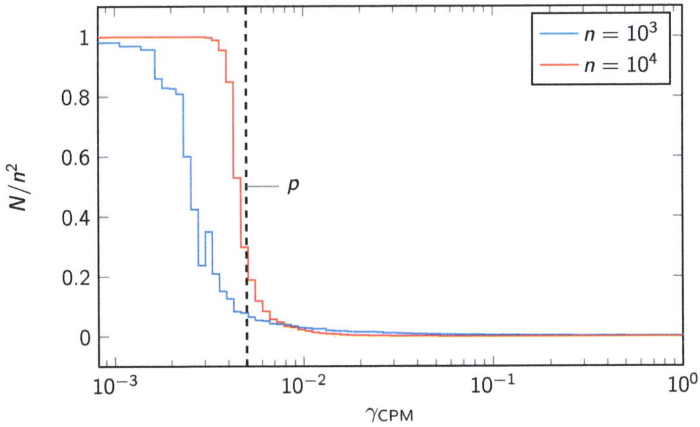

Fig. 4.2 Resolution profile for ER graph

Considering that $-(sn)^2 D(q, p)$ dominates both the upper and the lower bound, we can write that $\Pr(S(sn, q) \subseteq \mathcal{G}(n, p)) = e^{\Theta(-(sn)^2 D(q,p))}$, where $f = \Theta(g)$ is the asymptotic notation to state that f is asymptotically bounded below and above by g. Indeed, this provides the crucial insight into the asymptotic behaviour. For each $p \neq q$ the probability decays as a Gaussian, with a rate depending on the "distance" between p and q as expressed by the Kullback-Leibler divergence. Furthermore, the larger the proportional subgraph as expressed by s, the less likely a subgraph of different density than p can be found. Combining this idea for all communities, the probability for a partition σ with community sizes n_c and densities p_c should then scale as

$$\Pr(\sigma) = \exp\left(-\sum_c \binom{n_c}{2} D(p_c, p)\right). \tag{4.8}$$

Notice that for the two trivial partitions of all nodes in a single community, or every node in its own community, we obtain that $\Pr(\sigma) = 1$. This implies that such partitions are never significant, since they can always be found in any graph. We then define the significance as

$$\mathrm{Sig}(\sigma) = -\log \Pr(\sigma) = \sum_c \binom{n_c}{2} D(p_c, p) \tag{4.9}$$

for finding significant partitions. Ideally, a significant partition should have a low probability of appearing in a random graph, hence the significance $\mathrm{Sig}(\sigma)$ should be relatively high.

For benchmark networks, indeed this measure of significance works quite well, see Fig. 4.3. There, we report the same results as earlier (see p. 111), but we now

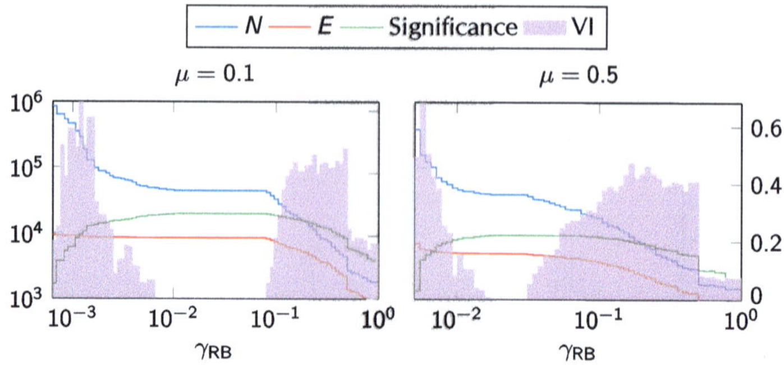

Fig. 4.3 Significance for benchmark graphs

include the significance results (referring to $\text{Sig}(\sigma)$). As we can see, the significance is maximal for the plateau at which we recover the "correct partition". Nonetheless, the measure indicates that slightly more refined partition (just right of the plateau) are also quite significant. Still, the measure of significance is actually maximal for the plateau.

For hierarchical benchmark graph results are similar, see Fig. 4.4. These networks have $n = 10^3$ nodes, and each node has a degree of $k_i = k = 20$. There are 10 large communities of 100 nodes each, and each large community consists of 5 smaller communities of 20 nodes each. There we observe two plateaus for $\mu_2 = 0.1$ (we have used $\mu_1 = 0.1$ for both results), corresponding to the two levels of the hierarchy. The significance of the more refined partition (the second level) is higher however, whereas the more broad partition (the first level) is less significant. This makes sense since the refined communities are very clearly defined, and so are very unlikely to be found in a random graph. The broader communities are also relatively clearly defined, but it contains a refinement that is less likely to be present in a random graph. For $\mu_2 = 0.5$ the two plateaus have merged into a single plateau, which is the most significant partition found. Again, this makes sense, since the smaller communities are much less clearly defined, while most links still fall within the larger community (since $\mu_1 = 0.1$). The maximal significance attains about 1.4×10^4 for $\mu_2 = 0.5$, while for $\mu_2 = 0.1$ the maximum is about 3.7×10^4. This suggest that the communities are more clearly define for $\mu_2 = 0.1$ then for $\mu_2 = 0.5$, as expected. Hence, the significance of a partition can be quite well used to find out what partitions are relatively significant at what resolution.

4.2.4 *Scanning for Significance*

We know how to scan the resolution parameter range without too much calculation by bisectioning. In addition, we have seen that the significance $\text{Sig}(\sigma)$ has the tendency to be maximal for some interesting partition. Hence, we might alter the bisectioning algorithm somewhat in order to look for the γ that maximizes the significance. So, in this case, we only use the significance to choose a particular value of γ that works

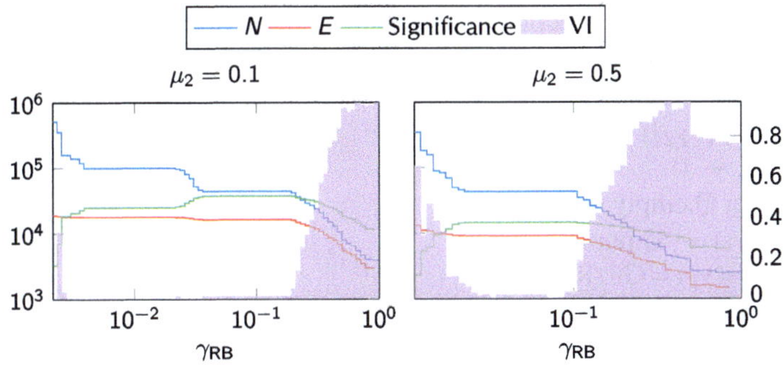

Fig. 4.4 Hierarchical scanning results

"well". Of course, this only returns a single value of γ, which in addition might be a local optimum. To obtain the full overview of the resolution profile, the scan has to be done over the whole range of γ, as in Sect. 4.1. An alternative approach would be to optimize significance itself, which we will consider in the next section.

By bisectioning, we know we don't have to scan some $\gamma \in [\gamma_1, \gamma_2]$ if $N(\gamma_1) = N(\gamma_2)$ for some γ_1 and γ_2. Hence, this reduces significantly the number of values of γ which we have to scan. If we are only interested in the γ for which $\text{Sig}(\sigma)$ is maximal, we can additional only scan those ranges for which the significance is maximal. However, recursion is inherently depth first. It tries to bisect as long as necessary some range, and only returns whenever it has reached some limit. In this case, it would be preferable to have a breadth first search, so as to cover a broad range of values. In order to do so, we will rely on a queue instead of recursing, see Algorithm 8.

4.2.5 Optimizing Significance

Scanning for a significant resolution, as in the previous section, provides us with a way to choose one of the partitions returned by CPM. However, it might be that there are other partitions, not revealed by CPM, that have a higher significance still. Hence, we might also try to optimize the significance directly. This comes down to using the significance as an objective function instead.

Notice that optimizing significance is no longer scale-invariant. After all, given a partition and a graph, pick a subgraph that consists of only a single community. Then the significance $\text{Sig}(\sigma)$ of that partition, defined on the subgraph equals 0, since $D(p_c, p) = 0$. This is also the case for all nodes as singleton communities, since then $\binom{n}{2} = 0$. Since this constitutes the minimum, it is unlikely that no other partition provides a higher significance. In particular, splitting the community in two smaller communities will in general give a non-zero significance. Hence, the same partition no longer remains optimal on all community induced subgraphs, and the method is hence not scale-invariant.

Algorithm 8 Recursive bisectioning maximizing the significance

function SIGRES(γ_1, γ_2, results)
  Q $\leftarrow$ empty queue
  Q.push([γ_1, γ_2])
  max_sig $\leftarrow$ 0
  while not Q.empty **do**
    [γ_1, γ_2] $\leftarrow$ Q.pop()
    results(γ_1) = GETRESULTS(γ_1)
    results(γ_2) = GETRESULTS(γ_2)
    dN $\leftarrow |N(\gamma_1) - N(\gamma_2)|$
    dG $\leftarrow |\gamma_1 - \gamma_2|$
    mS $\leftarrow \max(\text{sig}(\gamma_1), \text{sig}(\gamma_2)$
    if dN $> \epsilon$ and dG $> \delta$ and mS >= max_sig **then**
      max_sig $\leftarrow$ mS
      $\gamma_{\text{mid}} \leftarrow \frac{\gamma_1+\gamma_2}{2}$
      Q.push([$\gamma_1, \gamma_{\text{mid}}$])
      Q.push([$\gamma_{\text{mid}}, \gamma_2$])
    end if
  end while
end function

Optimizing significance is not too difficult. As before, we look at the difference of moving some node i from a community r to a community s. Let us assume that i has e_{ir} edges to community r and e_{is} edges to community s. The increase in significance is then

$$\begin{aligned}\Delta \text{Sig}(\sigma) = \binom{n_r}{2} D(q_r, p) - \binom{n_r - 1}{2} D(q'_r, p) \\ \binom{n_s}{2} D(q_s, p) - \binom{n_s + 1}{2} D(q'_s, p),\end{aligned}$$

where

$$\begin{aligned}q'_r &= \frac{m_r - e_{ir}}{\binom{n_r-1}{2}} \\ q'_s &= \frac{m_s + e_{is}}{\binom{n_s+1}{2}}.\end{aligned}$$

We can then perform the same greedy algorithm as before. However, if we also want to aggregate the graph, and then still correctly move communities, we again need the node size n_i, similar as for CPM. Suppose we have this node size, then moving a node with size n_i amounts to

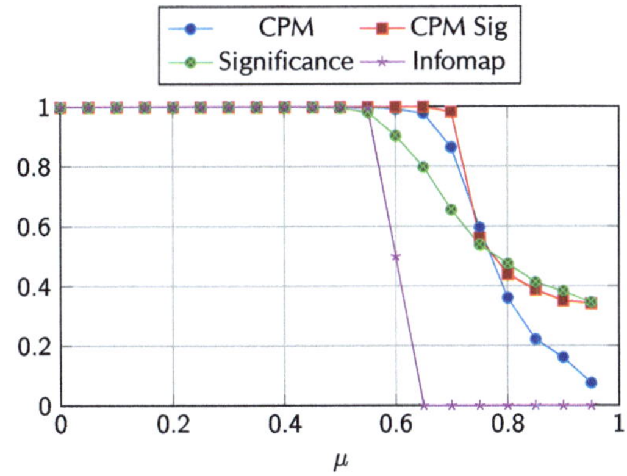

Fig. 4.5 Benchmark results for significance

$$\Delta\text{Sig}(\sigma) = \binom{n_r}{2}D(q_r, p) - \binom{n_r - n_i}{2}D(q_r', p)$$
$$\binom{n_s}{2}D(q_s, p) - \binom{n_s + n_i}{2}D(q_s', p),$$

where

$$q_r' = \frac{m_r - e_{ir}}{\binom{n_r - n_i}{2}}$$
$$q_s' = \frac{m_s + e_{is}}{\binom{n_s + n_i}{2}}.$$

Hence, we can use this to optimize significance using the Louvain algorithm (Algorithm 3, see p. 53), similar as we did for CPM.

The benchmark results are displayed in Fig. 4.5. It is clear that using significance to scan for the best γ parameter for CPM works quite well. In fact, is works better to scan for the best γ parameter then using our pre-calculated γ^* using information about the mixing parameter μ, as done in Sect. 2.4.3. Surprisingly however, optimizing significance itself results in a worse performance than scanning for the optimal γ parameter for CPM. This is presumably due to some local minima in which the significance optimization gets stuck, while this is not the case for CPM. In particular, it is likely that it will find denser subgraphs within the LFR communities, so that it doesn't find the actual communities. Nonetheless, optimizing significance works reasonably well, and outperforms Infomap in this case.

References

1. Aldecoa R, Marín I (2011) Deciphering network community structure by surprise. PLoS One 6(9):e24195. doi:10.1371/journal.pone.0024195
2. Aldecoa R, Marín I (2013) Surprise maximization reveals the community structure of complex networks. Sci Rep 3:1060. doi:10.1038/srep01060

3. Cover TM, Thomas JA (2012) Elements of information theory. Wiley, London. ISBN 1118585771
4. Delvenne JC, Yaliraki SN, Barahona M (2010) Stability of graph communities across time scales. In: Proceedings of the National Academy of Sciences of the United States of America 107(29):12755–12760. doi:10.1073/pnas.0903215107
5. Krings G (2012) Extraction of information from large networks. Ph.D. thesis, Université catholique de Louvain, Belgium
6. Mirshahvalad A, Lindholm J, Derlén M, Rosvall M (2012) Significant communities in large sparse networks. PloS one 7(3):e33721. doi:10.1371/journal.pone.0033721. arXiv: 1110.0305
7. Mucha PJ, Richardson T, Macon K, Porter MA, Onnela JP (2010) Community structure in time-dependent, multiscale, and multiplex networks. Science (NY) 328(5980):876–978. doi:10.1126/science.1184819. arXiv: 0911.1824
8. Reichardt J, Bornholdt S (2006) When are networks truly modular? Phys D Nonlinear Phenom 224(1–2):20–26. doi:10.1016/j.physd.2006.09.009
9. Ronhovde P, Nussinov Z (2009) Multiresolution community detection for megascale networks by information-based replica correlations. Phys Rev E 80(1):016109. doi:10.1103/PhysRevE.80.016109

Chapter 5
Modularity with Negative Links

UNTIL now we have constrained the weights on the graph to be positive ($w_{ij} > 0$). However, these weights might also be negative, a situation which comes quite natural when studying for example conflict. In these situations any animosity (e.g. war, fighting, conflict or distrust) can be represented by a negative link (some weight $w_{ij} < 0$) and the opposite (e.g. alliances or friendship) by a positive link (some weight $w_{ij} > 0$). Although the exact weight of course often plays some additional role, the distinction between positive and negative links is primal. Often we will simply consider negative links as having weight -1 and positive links having weight $+1$, although most concepts can be easily extended to weighted graphs. These type of networks are known as signed networks (or graphs) [6]. In this chapter we will analyse how this affects the proposed methods and offer some solutions.

5.1 Social Balance

The theory of social balance tries to explain the structure of positive and negative links in signed networks [1–4]. The idea is that whenever you are friends with somebody, you and your friend should have rather similar beliefs. Although friends perhaps tend to resemble each other more and more closely, it also works in the other direction: people tend to befriend those whom they share interests with (birds of a feather flock together). This latter process is known as homophily—like for the own kind. Enemies on the other hand should then be expected to think differently. We will elaborate on social balance in Chap. 7, and it will be the main focus in the second part.

The main result of social balance is that a graph that is balanced can be split into factions. Each faction corresponds to a set of nodes that is connected positively, while between factions there are only negative links. The number of factions is related to it being either weakly or strongly socially balanced. For a strongly socially balanced graph, it can be split into (at most) two factions, while a weakly socially balanced graph can be split into (possibly) more factions. More details can be found in Chap. 7.

V. Traag, *Algorithms and Dynamical Models for Communities and Reputation in Social Networks*, Springer Theses, DOI: 10.1007/978-3-319-06391-1_5,

The correspondence between factions and communities is clear, and we use the two words interchangeably in this chapter. However, instead of having relatively little links between communities, they should be negative.

5.1.1 Frustration

Of course, in reality we do not expect social balance to hold exactly, but only to some degree. A natural question therefore is whether it is possible to cluster a signed graph so that is has the least number of incorrect edges (i.e. positive link between factions or negative link within factions). At this point it is useful to introduce the negative and positive part of a signed graph. Let us denote by $G^- = (V, E^-)$ the negative graph and by $G^+ = (V, E^+)$ the positive graph, so that

$$E^- = \{(i,j) \in E \mid w_{ij} < 0\} \tag{5.1}$$

$$E^+ = \{(i,j) \in E \mid w_{ij} > 0\}. \tag{5.2}$$

The adjacency matrices A^+ and A^- are then defined accordingly, so that $A^+_{ij} = 1$ whenever $(i,j) \in E^+$ and zero otherwise, and similarly $A^-_{ij} = 1$ if $(i,j) \in E^-$ and zero otherwise. The original adjacency matrix is then $A = A^+ + A^-$. In addition, we will denote the signed adjacency matrix by $\tilde{A} = A^+ - A^-$, so that

$$\tilde{A}_{ij} = \begin{cases} -1 & \text{if } (i,j) \in E^- \\ 0 & \text{if } (i,j) \notin E \\ 1 & \text{if } (i,j) \in E^+ \end{cases} \tag{5.3}$$

We assume that $w_{ij} \neq 0$ whenever $(i,j) \in E$, so that there are no edges that have zero weight. We also define the positive and negative weights as $w^+_{ij} = \max\{w_{ij}, 0\}$ and $w^-_{ij} = \max\{-w_{ij}, 0\}$, so that if $w_{ij} > 0$ then $w^+_{ij} = w_{ij}$ and $w^-_{ij} = 0$ and if $w_{ij} < 0$ then $w^-_{ij} = -w_{ij}$ and $w^+_{ij} = 0$ and so $w^\pm_{ij} \geq 0$. In order to find factions such that there are the least number of violating edges, we need to minimize

$$\mathcal{H}_{\text{SB}} = \sum_{ij} \tilde{A}_{ij}(1 - \delta(\sigma_i, \sigma_j)).$$

Rewriting this, and removing those parts that do not depend on $\delta(\sigma_i, \sigma_j)$ this is equal to minimizing

$$\mathcal{H}_{\text{SB}} = -\sum_{ij} \tilde{A}_{ij}\delta(\sigma_i, \sigma_j) \tag{5.4}$$

Positive edges between communities and negative edges within communities are said to be frustrated. We also refer to Eq. 5.4 as the frustration, and we would like to minimize it.

Notice that if we are looking for only two communities, this reduces to bi-partitioning, which can effectively be done with the spectral method explained in Sect. 2.3.4. We defined a vector s such that $s_i = -1$ if node i is in community 1 and $s_i = 1$ if i is in community 2. If u is the eigenvector corresponding to the largest eigenvalue, then taking

$$s_i = \begin{cases} 1 & \text{if } u_i \geq 0 \\ -1 & \text{if } u_i < 0 \end{cases}$$

gives a reasonable partition in two groups. In particular, if the network is strongly socially balanced, so that it can be split exactly in two groups, this method will give an exact result (see Theorem 7.10). This can be seen as follows. Let x be a non-zero vector such that $x^\top x = 1$. Then

$$x^\top \tilde{A} x = \sum_{ij} x_i \tilde{A}_{ij} x_j.$$

Let $x_i x_j > 0$ if $\tilde{A}_{ij} = 1$ and $x_i x_j < 0$ if $\tilde{A}_{ij} = -1$, so that each $x_i \tilde{A}_{ij} x_j > 0$ since the graph is balanced. Suppose $u_j = x_j$ for all $j \neq i$ but that $u_i x_i < 0$. Then $u^\top \tilde{A} u < x^\top \tilde{A} x$ which contradicts the fact that u is the eigenvector corresponding to the maximal eigenvalue. Hence, indeed, if the graph can be partitioned into two groups, this split will be found exactly by the spectral bi-partitioning.

But imagine there are only a few negative links, and many positive links. According to this method, everything that is positively linked should be put in the same community. So, even though there might be communities that are well defined but only positively linked, they will be missed by using this method. So, this method might be too strict. In fact, the LP method has a similar problem (see Sect. 2.2.5 on p. 37). It essentially might put all nodes in the same community whenever they are positively connected.

Hence, although this frustration model would indeed minimize the number of frustrated links, it might not be exactly what we want. The method merges communities which are only relatively sparsely connected, while we might be interested in detecting separate communities.

5.2 Weighted Models

So far we have mostly discussed unweighed models, and simply stated that most models can be easily adapted for weighted networks. Perhaps the problem of negative links is quickly solved by simply allowing negative weights.

Let us first consider some of the weighted counterparts of the earlier models. We define $s_i = \sum_j w_{ij}$ as the strength of a node, as the weighted counterpart of the degree. Moreover, the total weight $w = \sum_{ij} w_{ij} = \frac{1}{2} \sum_i s_i$ is the weighted counterpart of the number of edges.

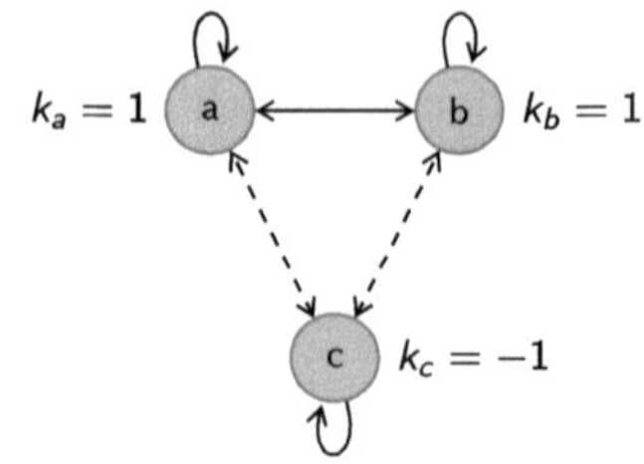

Fig. 5.1 Problem of modularity with negative links

Let us look at the weighted version of modularity for instance. We then arrive at

$$p_{ij} = \frac{s_i s_j}{2w}$$

for the expected weight of link (i, j), similar as before. The complete weighted form of modularity then becomes

$$\mathcal{Q} = \frac{1}{2w} \sum_{ij} \left(A_{ij} w_{ij} - \frac{s_i s_j}{2w} \right) \delta(\sigma_i, \sigma_j).$$

It is clear that quickly problems emerge if we allow $w_{ij} < 0$, since it might for example be that $w = 0$. Moreover, if $s_i < 0$ and $s_j > 0$, the expected weight $p_{ij} < 0$. In fact, even for a trivial example this does not work well. Consider the example provided in Fig. 5.1. The weighted degree of the three nodes a, b and c is $s_a = 1$, $s_b = 1$ and $s_c = -1$. The total weight is $w = \sum w_{ij} = 1$. The expected values $p_{ij} = s_i s_j / 2w$ equal the edge weights w_{ij}. Hence $w_{ij} - p_{ij} = 0$ for all links, and each possible community configuration results in a modularity $\mathcal{Q} = 0$, while the appropriate configuration is clear from the figure: a and b belong to the same community, and c to another community. Some modification to modularity is therefore required to detect communities in networks with (also) negative links.

Now let us take a look at the RB model, which for the configuration model gives largely similar issues as the modularity. But perhaps the ER null-model is less sensitive to issues of this kind. Let us define $p = w / \binom{n}{2}$. This amounts of course to a CPM with a rescaled resolution parameter, which is slightly easier to consider. The total weight inside a community is then given by $e_c = \sum_{ij} w_{ij} \delta(\sigma_i, c) \delta(\sigma_j, c)$, and the weighted density by $p_c = e_c / \binom{n_c}{2}$. In this case, communities should simply have a weighted density of $p_c > \gamma_{\mathrm{RB}} p$, while the density between communities c and d should be $p_{cd} < \gamma_{\mathrm{RB}} p$. Notice that for either null-model, using $\gamma_{\mathrm{RB}} = 0$ we arrive at the earlier model of frustration.

Let us try to repair the deficits of modularity in case there are negative links. In general we could write

$$\mathcal{H} = -\mathcal{H}^+ + \mathcal{H}^- \tag{5.5}$$

where $\mathcal{H}^+$ is the objective function defined on the network of positive links and $\mathcal{H}^-$ on the networks of negative links. The contribution for the negative links $\mathcal{H}^-$ is the opposite of that of positive links $\mathcal{H}^+$ since we want to minimize the number of negative links within a community instead of maximize them. Choosing different $\mathcal{H}^\pm$ leads to different community detection methods, similar as before, but adapted for when negative links are present.

Not all models necessarily have problems with negative weights, and let us briefly review which ones do and which don't. Let us first work out the RB model, which then becomes

$$\begin{aligned}\mathcal{H}_{\mathrm{RB}} = &-\sum_{ij}(w_{ij}^+ A_{ij}^+ - \gamma_{\mathrm{RB}}^+ p_{ij}^+)\delta(\sigma_i, \sigma_j) \\ &+\sum_{ij}(w_{ij}^- A_{ij}^- - \gamma_{\mathrm{RB}}^- p_{ij}^-)\delta(\sigma_i, \sigma_j).\end{aligned}$$

since $w_{ij} = w_{ij}^+ - w_{ij}^-$ and either $A_{ij}^+ = 1$ or $A_{ij}^- = 1$ we can simplify to

$$\mathcal{H}_{\mathrm{RB}} = -\sum_{ij}(w_{ij}A_{ij} - (\gamma_{\mathrm{RB}}^+ p_{ij}^+ - \gamma_{\mathrm{RB}}^- p_{ij}^-))\delta(\sigma_i, \sigma_j).$$

For the configuration null-model this gives

$$\mathcal{H}_{\mathrm{RB}} = -\sum_{ij}\left(w_{ij}A_{ij} - \left(\gamma_{\mathrm{RB}}^+\frac{k_i^+k_j^+}{2m^+} - \gamma_{\mathrm{RB}}^-\frac{k_i^-k_j^-}{2m^-}\right)\right)\delta(\sigma_i, \sigma_k). \tag{5.6}$$

So, when distinguishing negative and positive links, using the configuration null model is only equivalent to the original method when

$$\gamma_{\mathrm{RB}}^+\frac{k_i^+k_j^+}{2m^+} - \gamma_{\mathrm{RB}}^-\frac{k_i^-k_j^-}{2m^-} = \gamma_{\mathrm{RB}}\frac{(k_i^+ - k_i^-)(k_j^+ - k_j^-)}{2(m^+ - m^-)}$$

which in general will not be the case. However, when using the ER null-model, we arrive at

$$\mathcal{H}_{\mathrm{RB}} = -\sum_{ij}(w_{ij}A_{ij} - (\gamma_{\mathrm{RB}}^+ p^+ - \gamma_{\mathrm{RB}}^- p^-))\delta(\sigma_i, \sigma_j) \tag{5.7}$$

so that whenever $\gamma_{\mathrm{RB}}^+ p^+ - \gamma_{\mathrm{RB}}^- p^- = \gamma_{\mathrm{RB}} p$ or when $\gamma_{\mathrm{RB}} = \frac{\gamma_{\mathrm{RB}}^+ p^+ - \gamma_{\mathrm{RB}}^- p^-}{p}$ the original method is equivalent. Similarly for CPM, the original method is equivalent when $\gamma_{\mathrm{CPM}} = \gamma_{\mathrm{CPM}}^+ - \gamma_{\mathrm{CPM}}^-$.

One might wonder whether it is not simply an issue of shifting the adjacency matrix A by some constant in order to make it positive. Let us briefly reflect on this possibility. We thus have the following shifted weighted adjacency matrix $A' = A + c$

for some constant $c \geq 0$ such that $A'_{ij} \geq 0$ for all ij. We already saw that modularity has some issues with negative weights, so maybe they've disappeared when simply shifting the matrix. Indeed the example on Fig. 5.1 can be trivially repaired for modularity by simply shifting the matrix with a constant $c = 1$. However, in general, this way of repairing things does not coincide with our solution in Eq. 5.5. However, for CPM (and so by extension RN and the RB method with the ER null model), this simply corresponds to a shift in the resolution parameter γ_{CPM}, which is not the case for modularity. Arguably, if a partition is optimal when shifting the matrix, but not according to Eq. 5.5, it does not seem to constitute a good partition. Nonetheless, we might wonder if there always exists some c such that the partition is optimal for both modularity on a shifted matrix and for Eq. 5.5. This remains an open question, but given the problems of modularity described in Chap. 3, it is doubtful. At any rate, modularity is not simply shift invariant in this sense, whereas CPM is shift invariant (up to a concomitant shift in the resolution parameter).

5.3 Implementation and Benchmark

The implementation of the negative links is not too difficult in most algorithms reviewed in Sect. 2.3. We already briefly saw the application of the spectral bisectioning, which is easily applied to any matrix. For the other algorithms we have to make a small change however.

Let us review the greedy method, which forms the core of the Louvain method, for negative links. Let us take the RB model with the configuration null-model as an example. The change when moving node i from community c to community d is in general then

$$\begin{aligned}\Delta\mathcal{H}(\sigma_i = c \mapsto d) = &\left[(e^+_{id} - \gamma\langle e^+_{id}\rangle_{p_{ij}}) - (e^-_{id} - \gamma\langle e^-_{id}\rangle_{p_{ij}})\right] \\ &- \left[(e^+_{ic} - \gamma\langle e^+_{ic}\rangle_{p_{ij}}) - (e^-_{ic} - \gamma\langle e^-_{ic}\rangle_{p_{ij}})\right].\end{aligned}$$

Earlier however, it was clear that a community needed to be connected, and so it made sense to only consider the communities of neighbours. This is no longer the case unfortunately when introducing negative links. After all, suppose for example that $k^+_i = 0$, node i has only negative links. In that case we seek to minimize $e^-_{ic} - \gamma\langle e^-_{ic}\rangle_{p_{ij}}$ which probably happens when $e^-_{ic} = 0$. Hence, when negative weights are included, we are obliged to loop through all communities, not only the communities of neighbours. So the greedy method should be only slightly adapted, and is displayed in Algorithm 9.

Another small change in the implementation that makes it slightly easier to work with negative links is to work with layers. This can then also be easily extended to work with more complicated graphs, with multiple types of links, or multiple types of models for different layers. In general, the idea is to have different graphs $G_1 = (V, E_1), G_2 = (V, E_2), \ldots$, with the same nodes in it, and that we calculate the cost by summing the objective function for these different layers. Each layer

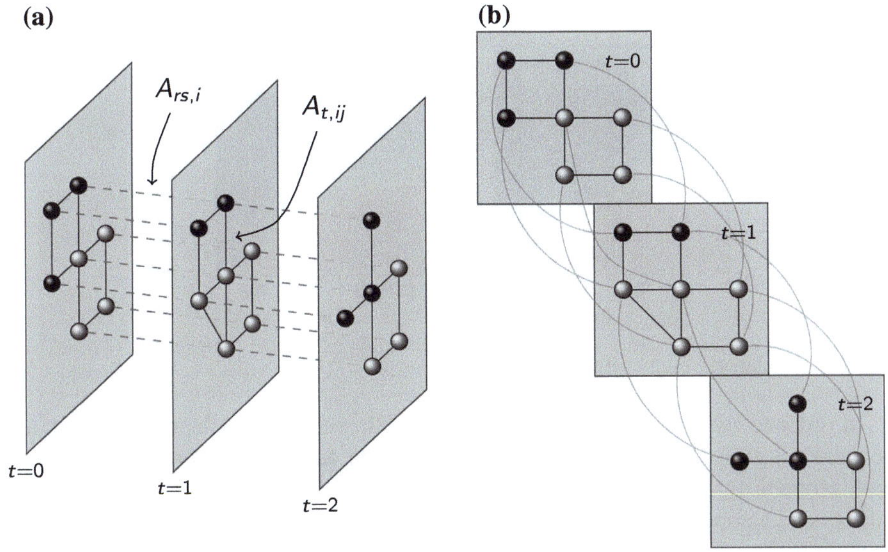

Fig. 5.2 Multi-slice modularity to layers. **a** Slices, **b** layers

Algorithm 9 Greedy method for negative links/multiple layers

```
function GREEDY(Graph G)
    initialize σ_i ← i for all nodes i
    while improvement do
        for all nodes i do
            for all communities c do                        ▷ All communities
                Δ_c ← Σ_ℓ ΔH_ℓ(σ_i = r ↦ c)                 ▷ Sum over all layers
            end for
            σ_i ← arg max_c Δ_c                             ▷ Greedily, maximum choice
        end for
    end while
    return σ
end function
```

only contains positive weights. So if the original graph G contains both positive and negative links, we split it in a positive part G^+ and G^- as stated earlier, which represent our layers G_1 and G_2, and we keep track of which layer contains the negative links, so that we try to maximize that objective function, instead of minimize it. If we have a dynamic graph so that at different times there are different edges present/absent, we can use a similar technique. Normally some links between graphs at consecutive times are added, so as to obtain a dynamic view of the partition [5]. In that way, we can keep most of the original implementation details. Specifically, for the Louvain method, this way of implementing allows to keep the same functions for aggregating the graph (layers) as before (see Fig. 5.2 for an illustration of this idea).

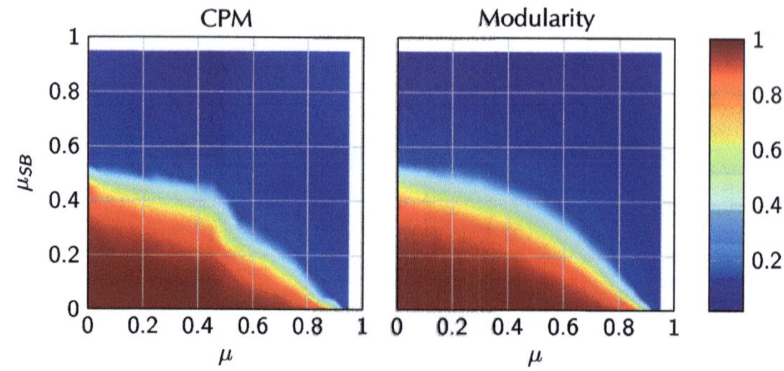

Fig. 5.3 Benchmark results with negative links

In order to see if a method is performing well, we need to adapt the benchmark networks slightly. We can do so by first generating an ordinary benchmark, so that with probability μ links fall within a community, and with probability $1-\mu$ they fall outside a community. We introduce also the mixing parameter μ_{SB}, so that with probability μ_{SB} a link within a community is positive, and with probability $1-\mu_{SB}$ such a link is negative. Similarly for links outside a community, with probability μ_{SB} this is a negative link, while with probability $1-\mu_{SB}$ this is a positive link. So, for $\mu_{SB}=0$ all links within communities are positive and all links between communities negative.

We can repeat again the same analysis we did before on when the communities are well defined. For the different densities we obtain

$$p^+_{\text{in}} = \frac{(1-\mu)(1-\mu_{SB})\langle k\rangle}{n_c-1} \qquad p^-_{\text{in}} = \frac{(1-\mu)\mu_{SB}\langle k\rangle}{n_c-1}$$

$$p^+_{\text{out}} = \frac{\mu\mu_{SB}\langle k\rangle}{n-n_c} \qquad p^-_{\text{out}} = \frac{\mu(1-\mu_{SB})\langle k\rangle}{n-n_c}$$

so that if $\mu_{SB} < 1/2$, we obtain that communities are well defined as long as

$$\mu < \frac{n-2n_c+1}{n-1}.$$

Surprisingly however, if $\mu_{SB} > 1/2$ the communities are well defined if

$$\mu > \frac{n-2n_c+1}{n-1}.$$

This is due to the effect that there are relatively many negative links within a community if $\mu_{SB} > 1/2$. Effectively there is a phase transition at $\mu_{SB} = 1/2$ so that quite suddenly, the regime where the communities are well defined changes.

The benchmark results are displayed in Fig. 5.3 for CPM and modularity with negative links. We use the parameters of $\gamma^{\pm}_{RB} = 1$ (i.e. modularity) and $\gamma^{\pm}_{CPM} = p^{\pm}$, where $p^{\pm}$ is the average positive/negative density within a community. The results are reasonably similar, and both work quite well until $\mu^- < 0.5$, and as can be expected if becomes increasingly more difficult for higher μ^+. For $\mu^- > 0.5$ neither CPM nor modularity is able to recover the planted partition correctly. CPM does seem to perform a little bit better for high μ_{SB} and high μ than modularity however.

References

1. Cartwright D, Harary F (1956) Structural balance: a generalization of Heider's theory. Psychol Rev 63(5):277–293. doi:10.1037/h0046049
2. Cartwright D, Harary F (1968) On the coloring of signed graphs. Elem Math 23(4):85–89. doi:10.5169/seals-26032
3. Cartwright D, Harary F (1979) Balance and clusterability: an overview. Academic Press, New York
4. Harary F (1953) On the notion of balance of a signed graph. Mich Math J 2(2):143–146. doi:10.1307/mmj/1028989917
5. Mucha PJ, Richardson T, Macon K, Porter MA, Onnela JP (2010) Community structure in time-dependent, multiscale, and multiplex networks. Science 328(5980):876–878. doi:10.1126/science.1184819. arXiv:0911.1824
6. Zaslavsky T (1982) Signed graphs. Discrete Appl Math 4(1):47–74. doi:10.1016/0166-218X(82)90033-6

Chapter 6
Applications

In this section we will see two applications of community detection. The first focuses on conflict in international relations, and tries to determine the influence of trading communities on the probability of conflict. The second focuses on citation networks and the effect of negative links in networks.

6.1 Communities in International Relations

We will investigate international relations, which have both positive as well as negative components. Negative links are operationalised as conflict, while positive links are represented by alliances (for example a defence pact).

When in the early 1990s the Communist bloc fell apart, many wondered what type of world would lie ahead. Two broad scenarios were sketched. On the one hand, democracy was seen as the sole surviving ideology, and conflict was expected to diminish, and Francis Fukuyama proclaimed "the end of history" [16]. On the other hand, conflict was no longer fuelled by ideological considerations and Samuel Huntington argued that conflict would simply run across different lines, namely civilizations, in his book entitled *The Clash of Civilizations* [28]. Clearly it would be interesting to see to what extent communities of international relations would correspond to what scenario.

To that end, we analyse international relations taken from the Correlates of War [19, 20] data set over the period 1993–2001, where military alliances can be represented by positive links and conflicts by negative links. The data set contains a wide variety of disputes, for example border tensions between Colombia and Venezuela, the deployment of Chinese submarines to Japanese islands, and Turkish groups entering Iraqi territory. Disputes were assigned hostility levels, from "no militarized action" to "interstate war," and we chose the mean level of hostility between two countries over the given time interval as the weight of their negative link. The alliances we coded one of three values, for (1) entente, (2) non-aggression pact, or (3)

V. Traag, *Algorithms and Dynamical Models for Communities and Reputation in Social Networks*, Springer Theses, DOI: 10.1007/978-3-319-06391-1_6,

defence pact. The disputes $w_{ij}^-(t)$ and alliances $w_{ij}^+(t)$ are both normalized to values in the interval $w_{ij}^{\pm}(t) \in [0, 1]$ for each year t. They bear equal weight in the overall link value $w_{ij}(t) = w_{ij}^+(t) - w_{ij}^-(t)$, and the final weight is $w_{ij} = \frac{1}{T}\sum_t w_{ij}(t)$ with T the total number of years included. For example, if two countries have a defence pact for a single year, the weight $w_{ij} = 1/T$, while if they have had war for 3 years (and no other alliances or conflicts), $w_{ij} = -3/T$. The largest connected component consists of 161 nodes (countries) and 2,517 links (conflicts and alliances).

The result of the analysis using the RB model with configuration null model adapted for negative links is shown in Fig. 6.1. Countries of the same colour belong to the same community, which in this context is perhaps more appropriately labelled a *power bloc*. The power blocs can be identified as follows: (1) the West; (2) Latin America; (3) Muslim World; (4) Asia; (5) West Africa; and, (6) Central Africa. If we detect communities by using only positive links, there is an agreement of about 64 % with the configuration in Fig. 6.1, while if using only negative links, there is an agreement of about 30 %, measured using NMI.

This resembles quite closely the configuration sketched by Huntington [28], with a few notable exceptions. The West African power bloc is an additional insight that is absent in Huntington's configuration. A major difference with Huntington is that China itself does not constitute a separate bloc, nor does Japan or India. Some other noteworthy differences are Pakistan and Iran which are grouped with the West, while South Korea and South Africa are grouped with the Muslim World.

If we run the algorithm with $\gamma_{\text{RB}}^+ = 0.1$ and $\gamma_{\text{RB}}^- = 1$, North America merges with Latin America, while Europe becomes an independent community, and North Africa and the Middle East align with Russia and China. When setting $\gamma_{\text{RB}}^+ = 1$ and $\gamma_{\text{RB}}^- = 2$, in contrast, former Soviet countries separate from Russia and form an independent community. Using a range of values for $\gamma_{\text{RB}}^{\pm}$, one can detect various levels in the community structure.

These results do not imply that conflicts take place between power blocs only, as 24 % of all conflicts actually take place *within* blocs. For example, Georgia and Russia had serious conflicts, and DR Congo and Rwanda had theirs, but each of these pairs is grouped together nevertheless. In these cases, the alliances overcame the conflicts in the grouping, confirming that a configuration of international relations is more than the sum of bilateral links.

In sum, although Huntington's configuration of civilizations was questioned [26, 53], it seems to be fairly robust and with some marked exceptions is confirmed by our analysis. However, this does not imply that this is only influenced by civilizations, since many other underlying factors may play a role. In fact, the more interesting question is how such a structure comes about. Many theories could be relevant, including the democratic peace theory [12, 27], which predicts few conflicts between democratic countries but fails to predict that in actuality, most conflicts occur between democratic and non-democratic countries; the realist school [31], which emphasizes geopolitical concerns; and, finally, the trade-conflict theory [50], which argues that (strong) trade relations diminish the probability of a dispute, or lower its intensity. We will investigate this in in the remainder of this section, with a particular focus on the effect of trade on conflict.

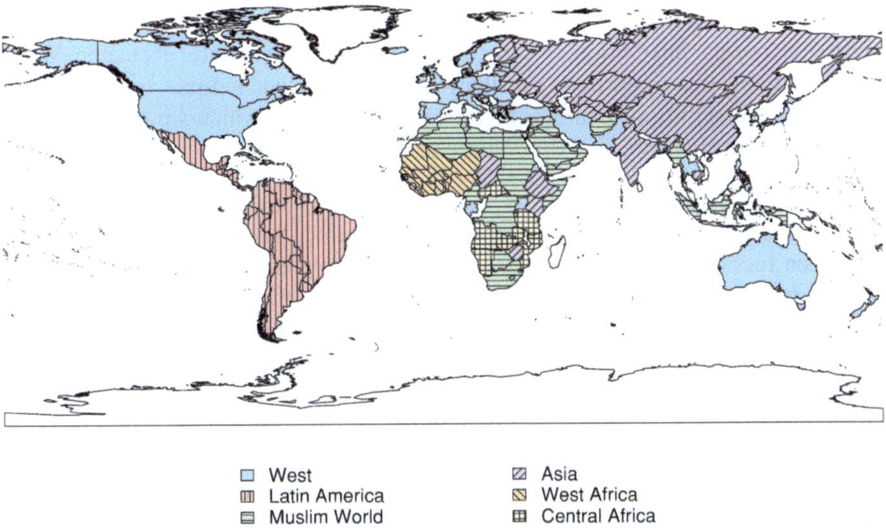

Fig. 6.1 Communities in the conflict and alliance network

6.1.1 Direct Trade and Conflict

Theories about the relationship between trade and conflict have a long tradition in international relations scholarship. Most of these focus on bilateral relationships, explaining whether and how increased levels of trade between two states affect their probability of direct conflict. This focus on only the links involving the two states is known as a dyadic analysis, where a dyad simply refers to the two nodes (states in this case), and the relationship between them. A smaller body of work also examines the ways in which dyadic dependence affects the probability of systemic conflict, although the findings from this work remain tentative [43, 45]. Recent work has begun using network analytic measures to demonstrate that indirect trade relations also have important effects on interstate conflict [5, 14, 35, 36].

A significant limitation of the existing literature is its almost exclusive focus on direct trading relationships. Analysing only dyadic trade relations over-simplifies the complexity of interdependence and, as a result, loses sight of the ways in which trade reduces conflict even among states that trade very little with each other. We argue that indirect trade dependence creates significant costs of conflict in addition to those created by the levels of direct trade between states. In addition, the conflict-reducing effects of interstate trade are heightened within trading communities.

Several studies have provided evidence that indirect trade relations reduce conflict. The probability of conflict is lower among dyads with more trading partners in common [14, 36] and among dyads that are generally more well-connected to other states in the trade network [14]. Yet this literature has not fully explained the causal mechanisms underlying these effects. Dorussen and Ward [14] argue that the key mechanism at work here is informational: trade decreases the likelihood of conflict by facilitating regular interaction, informational exchange and cultural exchange. While acknowledging this important contribution, we argue that indirect trade relations reduce the probability of conflict in two additional ways, which we refer to as the "Combatant Mechanism" and the "Non-combatant Mechanism."

The Combatant Mechanism

That trade between potential combatants may affect their incentives to fight has long been recognized, yet we argue that these incentives may also be affected by their trade relationships with other states. We build on the opportunity-cost theory of interdependence. Traditional formulations of this argument focus on the extent to which the potential participants in a conflict stand to have their trade with each other interrupted or otherwise adversely affected [3, 44, 46, 49]. Thus far, the opportunity-cost model has focused on the potential effects of conflict between a pair of states on their trade with each other. If a pair of states trades with each other relatively little (or not at all), this theory would predict that trade would have little effect on the probability of war between them.

Yet in a world of a complex trade network conflict may also interrupt trade flows other than those between the potential combatants. Although a pair of states may not have a trade relationship with each other, they would jeopardize their trading relationships with other states by going to war, and therefore have a disincentive to do so. Entering a conflict could interrupt a state's trade with states not involved in the conflict in various ways. Trade relations are highly interdependent, so the terms of trade within any pair of states depend on the terms of trade they have with other states [1, 2]. A warring state may divert resources previously used to produce certain exports in order to facilitate war-time production, thus reducing or cutting off those export flows. Conflict could result in decreased demand for the state's exports to the extent that demand is dependent on other trade flows interrupted by the conflict. Conflict may interrupt the supply of imports to the state to the extent it affects the supply chain for those imports. Finally, even when states do not directly trade with each other, indirect trade dependence increases the opportunity cost of a potential conflict between them because the uncertainty associated with war may cause their trading partners to seek other, more stable markets or suppliers [11, 18, 36, 49]. In summary, the combatants themselves might incur a cost, even though they do not trade directly.

The Non-combatant Mechanism

Indirect trade dependence also reduces the likelihood of conflict in a second way that has been under-theorized in the trade-and-conflict literature. Conflict creates costs for states that are not involved in it, but that are dependent on trade relations with the warring states. By interrupting trade flows, conflicts create negative externalities for non-participant states, including by decreasing their access to commerce, increasing the costs of their imports and decreasing the demand for their exports. As a result, indirect trade dependence reduces the probability of conflict by increasing the incentives for third parties to attempt to prevent the conflict (for related arguments, see [5, 6, 13]). While many states do not have the capability to significantly influence the potential combatants, others can and do use their power to deter wars that would damage international commerce [21, 30, 32]. If the non-combatants are too diffuse, diverse or numerous, problems of collective action might preclude an intervention [41]. However, if these states are sufficiently interdependent themselves, and not too diverse, they may overcome problems of collective action, and intervene. In summary, even countries not directly involved in the conflict can be affected by the conflict through indirect trade networks, and so have some interest in preventing it.

6.1.2 Trading Communities and Conflict

In which situations are the disruptions to trade caused by conflict most likely to create the types of costs that, in turn, reduce the probability of conflict? In addition, how do we systematically account for the ways in which indirect, networked trade relations affect conflict behaviour? Dorussen and Ward [14] propose that we can systematically capture the effects of indirect trade links by using the concept of maxflow, particularly because it may be a good proxy for the information flow between the members of a dyad that is facilitated by their trading relations. The maxflow can be defined as follows. Let r and s be two nodes, which play the role of source and sink. We seek to maximize the total flow $\sum_i f_{is}$ towards s, where f_{is} is the flow from node i to node s, under the constraint that each flow does not exceed its capacity w_{ij}, and the flow into a node equals the flow from a node $\sum_j f_{ij} = \sum_j f_{ji}$. In total then the maxflow is defined by

$$
\begin{aligned}
&\max \sum_i f_{is} \quad \text{such that} \\
\sum_j f_{ij} &= \sum_j f_{ji} \quad \text{for all} \quad i \neq r, s \\
&f_{ij} \leq w_{ij}.
\end{aligned}
$$

The maxflow is the same as the minimum cut between two nodes, and so equal to the number of independent paths if each link has unit capacity. The maxflow is thus a useful concept for understanding the effects of the informational mechanism proposed by Dorussen and Ward [14].

Yet this concept does not capture other ways in which the networked structure of international trade may be relevant to the mechanisms we propose. We illustrate this point using the stylized exchange networks provided in Fig. 6.2 where each edges has unit capacity. In network (a), the maxflow between nodes 1 and 2 is equal to 4 because a connection can be made between 1 and 2 using 4 possible independent routes: 1–3–2, 1–4–2, 1–5–2, and 1–6–2. In network (b), the maxflow between nodes 1 and 2 is also equal to 4. The additional flows in network (b) do not provide additional

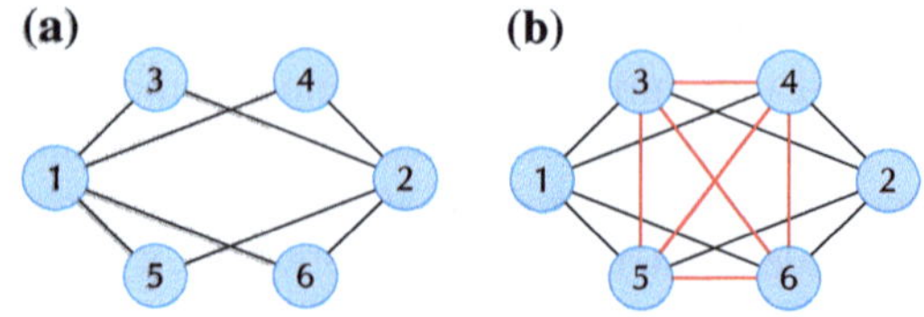

Fig. 6.2 Two trade networks of different densities. **a** Sparse. **b** Dense

possible independent paths between 1 and 2. Thus, a theory based on the concept of maxflow would make equivalent predictions regarding the extent to which indirect trade links between 1 and 2 would affect their conflict propensity in the two networks. Yet the two networks vary in terms of density: network (b) is significantly more densely connected than network (a). In terms of trade flows, network (b) can be thought of as more highly interdependent than network (a).

We argue that this difference between the two groups is crucial. In a highly interdependent group, when individual trade flows are cut off by conflict among the group's members, the probability that this will adversely affect other flows is higher. Therefore, the costs of a conflict involving two members of such a group would be especially high. Preventing such a conflict may be difficult and costly itself, but the group's members will have particularly important incentives to overcome this collective action problem. By contrast, when the potential combatants are not embedded within a single group of highly interdependent states, fewer flows may be interrupted by the conflict, and thus the economic costs of the conflict would be significantly lower, everything else equal.

This argument points to the concept of communities. Trade creates groups of states at the sub-global level in which the effects of indirect trade dependence are especially significant. Within these trading communities, states have many trading partners in common and, therefore, their dependence on each other is often far greater than their dyadic trade levels would suggest. Some dyads within a trading community trade significantly with each other, such as two developed states that trade differing manufactured goods they specialize in producing. Other dyads within a trading community may trade directly very little, however. This can occur, for example, when two states are at opposite ends of a single supply chain. Another example is of two states that are individually dependent on exporting and importing similar goods to and from the same third country.

The key factors that have shaped the structure of the global trading network are also responsible for the formation of trading communities. Trade flows highly unevenly across the international system, which is not at all surprising when taking economic factors into account [23]. Geographic distance creates transaction costs that promote trade among close neighbours [48]. This suggests that trading communities may have a strong regional component, although this may not always be the case. A state with a highly specialized production capability may be in the same trading community with a distant state that has a complementary demand for that specialized good. More generally, we would not always expect that a group of geographically clustered underdeveloped states would be in the same trading community. We can expect such states to trade relatively little with each other. Thus, if they export to and import from differing markets, they are likely to be in differing trading communities. Africa is a prime example. It would indeed be surprising to find that Africa consists of a cohesive trading community given that most states in the continent are poor and sell many of their raw goods to richer states outside the continent. Historical factors likely also contribute significantly to the formation of trading communities. We would expect, for example, that colonization and decolonization have had significant impacts on the structure of world trade, with former colonies continuing to trade significantly with their former colonizers. More recently, many of the trading relations established within the Soviet Union have continued among the former Soviet states, and thus we would expect a significant likelihood that these states are in the same trading community.

These arguments lead to the principal hypothesis that the probability of conflict is lower between state dyads that are members of the same trading community.

6.1.3 The Trade Network

The first step in testing our hypothesis is to construct the international trade network, which we do by using the data provided by Gleditsch [22]. Constructing this network requires us to assign weights to the dyadic links between states, which we do using the trade flows between them. Specifically, we define these weights using the formula for dyadic trade dependence provided by Oneal and Russett [42] and used by much of the literature on which we build:

$$w_{t,ij} = \frac{x_{t,ij} + x_{t,ji}}{GDP_{t,i}}, \tag{6.1}$$

where $x_{t,ij}$ is the total exports from country i to country j in year t (which equals the total imports to country j from country i), and $GDP_{t,i}$ is the total GDP of country i for year t.

We maximized the RB model with the configuration null model to define trading communities at an annual basis over a range of resolutions. Figure 6.3 shows representative partitions for the year 2000 using resolutions levels that yield 3, 7 and 14 trading communities. At a relatively low resolution level, we observe 3 large trading communities. One community includes the bulk of the Western Hemisphere in what appears to be a US-centric community. In 2000, Argentina significantly devalued its currency, causing short-term changes in its trade relations. In previous years, Argentina was a member of the Western Hemisphere trading community. A few states outside the Americas are also members of this trading community, notably the United Kingdom and Israel, a finding likely driven by their close trade links with the United States. Others, such as Norway, Iceland and Ireland, have less significant trade links with the United States, but do have close links with the United Kingdom, suggesting that they are in this community largely because the

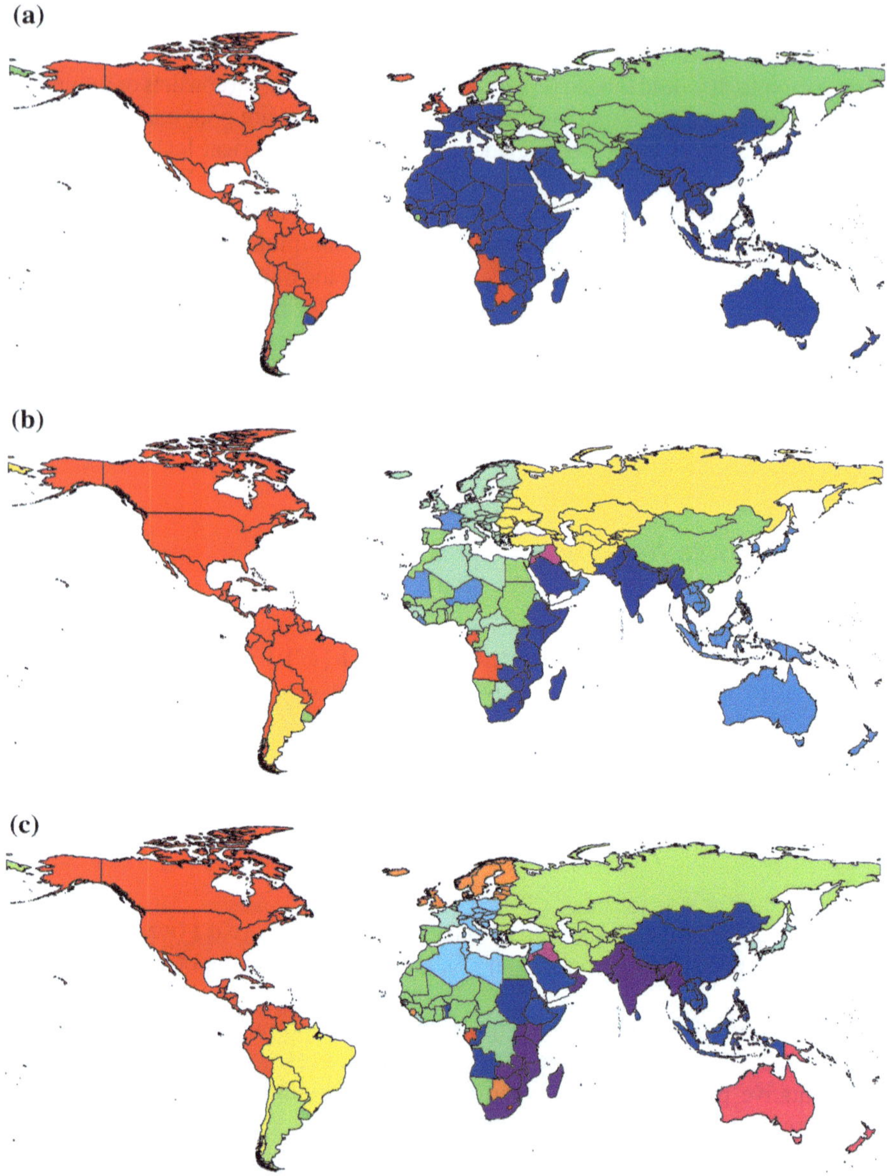

Fig. 6.3 Trading Communities in 2000. **a** Low resolution ($\gamma_{RB} = 0.6$). **b** Medium resolution ($\gamma_{RB} = 1.1$). **c** High resolution ($\gamma_{RB} = 1.7$)

United Kingdom is also. The second large community we see at this resolution level includes the former Soviet Union, Eastern Europe and parts of the Middle East. Finally, the rest of the world belongs to a trading community that includes Japan,

China, India, much of Europe, South-East Asia and most of Africa. This is arguably the most surprising among the findings at this level of resolution because it includes several major economies that are geographically dispersed. The surprising nature of this result suggests, in fact, that the trading community defined at this resolution level is actually an amalgamation of several sub-communities.

As the resolution level increases, so does the number of trading communities defined. Figure 6.3b shows a partition with seven communities. The Western Hemisphere community remains largely intact, which is not surprising given the level of dependence of most of these states on U.S. trade, and vice versa. Nonetheless, at this resolution level, the United Kingdom is no longer part of the Western Hemisphere community, and instead belongs to a smaller community consisting of Northern and Central Europe along with several of their African trading partners. We noted above that countries such as Norway, Iceland and Ireland were likely only defined as being in the Western Hemisphere community by virtue of their trade with the United Kingdom, so it is not surprising to observe that they "follow" the United Kingdom into this smaller community. Other states, such as Sweden, Finland and Denmark, are now defined as being in this community despite having previously been defined as part of the larger Russia-centred community rather than the Western Hemisphere community. This suggests that, at the lower resolution level, these countries are borderline cases; indeed, several other partitions at the low resolution level include these in the same community as the United States and United Kingdom. Aside from these states, the community of former Soviet Bloc states remains whole at this resolution level. The only former members of the Soviet Union not in the latter community are the Baltic states, a finding that is not surprising given that these economies have distanced themselves from Russia more so than any others. The largest community found in the low resolution level breaks into several communities at this level. The most notable of these are trading communities that include (1) China and many of its smaller trading partners; (2) South-East Asia, Australia and Japan; and (3) many states bordering the Indian Ocean, including South Asia and East Africa.

Finally, at a higher level of resolution, we observe several new trading communities. Three changes relative to the medium resolution are worth noting. First, Northern Europe and Central Europe seem to have split into two communities at high resolution. Second, several states in South America, most notably Brazil, form a sub-community within the larger Western Hemisphere community. Finally, Australia, New Zealand and several of the Pacific Island states have separated from South-East Asia into a smaller trading community that is most likely driven by Australian trade links. A large proportion of global trade is conducted within trading communities.

At a low level of resolution, most global trade has been conducted within the trading communities. Interestingly, the percentage of global trade conducted within these large communities decreased from about 90 % in 1960 to about 55 % in 2000, which suggests that globalization may have evened out global trade flows to a significant extent. At the medium level of resolution, about 40 % of trade is conducted within trading communities, despite the fact that only approximately 20 % of dyads are members of the same trading communities. This result means that a disproportionately large percentage of global trade is conducted within these groups. Finally, at a high level of resolution, approximately 30 % of trade is conducted within the 15 % of dyads that are joint members of these small subgroups.

6.1.4 Results

To test our hypothesis, we first create a variable that indicates whether, in a particular year, both members of a dyad were members of the same trading community (SAME TRADING COMMUNITY). As noted in Sect. 2.3, the modularity maximization algorithm may produce slightly different results each time it is run at a given resolution because there are many local maxima at which modularity is optimized. We could certainly choose one that appeared to have high face validity and test our hypothesis using it, but the validity of our results would depend on the validity of that particular partition. Instead, we use a construction that takes advantage of this feature of modularity maximization. For each resolution level, we run the modularity maximization algorithm 100 times. In each partition, we recognize that there is a certain degree of uncertainty regarding whether states have been correctly classified into trading communities. By running the algorithm many times for each dyad-year, we then code SAME TRADING COMMUNITY as "1" if it appears in the same trading community in more than 50 % of the partitions and "0" if it does not. Essentially this is a rounded consensus matrix.

Control Variables

As stated earlier, many variables are known to affect the propensity for conflict. We include these variables to show the effect of trade communities goes beyond that. We use Zeev Maoz's construction of dyadic militarized interstate disputes (MAOZMID) as the dependent variable [24, 29]. We coded the variable as "1" for dyad-years in which there was an onset of a militarized interstate dispute in which force was threatened or used, and "0" otherwise. We modified the coding of MAOZMID such that it indicates whether a MID was initiated in the year following the year in question, which has the same effect as lagging all of the independent variables by 1 year. Because we argue that SAME TRADING COMMUNITY should have a negative, significant relationship with MAOZMID regardless of the level of direct trade dependence, we control for this (DYADIC TRADE DEPENDENCE LOW) using the same formula in Eq. 6.1 used to calculate the weights in the trade network [42]. We include this control in Model 1 and remove it in Model 2 to demonstrate that our primary result is robust to the inclusion and exclusion of this measure. We also control for the maxflow (MAXFLOW), to capture some of the indirect effects informational mechanisms may have on conflict propensity. If SAME TRADING COMMUNITY has a significant relationship with MAOZMID despite the inclusion of these controls, this would provide evidence that the clustered structure of the trade network has an important relationship with conflict in ways not previously understood.

We also include several other controls that may affect the propensity for dyadic conflict and that have been used in much of the trade-and-conflict literature [14, 17, 42]. Democratic peace theorists argue that democracies have a lower propensity for conflict, especially with each other [7, 8, 15, 37, 38, 59]. We therefore control for the lower (DEMOCRACY LOW) and higher (DEMOCRACY HIGH) democracy scores in the dyad using the Polity IV data [39]. Shared membership in inter-governmental organizations (IGOs) may reduce the probability of conflict [13, 52], so we control for the number of shared IGOs memberships in the dyad using the Correlates of War 2 International Governmental Organizations Data [47]. Economic development may affect conflict propensity, so we follow Gartzke [17] and others in controlling for the lower of the GDP levels in the dyad-year. We also control for the effects of monadic power on the probability of conflict. The most powerful states are more actively engaged in interstate relations and may therefore be more likely to fight wars. We therefore include a dichotomous variable (MAJOR POWER) coded "1" for dyads in which at least one member is one of the five post-World War II major powers (i.e., United States, USSR/Russia, United Kingdom, China, and France). Allied states may be less likely to fight each other, so we include a dichotomous variable (ALLIANCE) coded "1" for dyads that have concluded an entente, neutrality pact or defence pact based on the Correlates of War (COW) Alliance Data Set [55, 56]. States may be more likely to attack weaker opponents. We therefore control for the natural logarithm of the ratio of the stronger state's COW capabilities index (CINC) to that of the weaker state (CAPABILITY RATIO).

We control for several geographic factors known to affect the propensity of dyadic conflict. Including geographic controls allows us to conduct a particularly strict test of the relationship between trading communities and conflict given that we know trading communities are clustered geographically. Wars are generally less costly for states to conduct against their immediate neighbours, so we construct a dichotomous variable coded "1" for dyads that share a land border or that are separated by less than 150 miles of water (CONTIGUITY). We also include a control measuring the natural logarithm of the distance between national capitals (DISTANCE). We adopt the method of Beck et al. [4] of including temporal spline variables and a measure of the duration of dyadic peace (PEACEYEARS) to control for duration dependence. Dyad-years with ongoing MIDs are excluded to avoid address problems of serial correlation. Our analysis includes the years 1960–2000.

Regression Analysis

Using this model, we tested our hypothesis over a large range of community detection resolutions. Table 6.1 provides the results of these models for resolutions yielding 3 (Low), 7 (Medium) and 14 (High) trading communities. The results provide substantial support for our hypothesis. States that are members of the same trading community are less likely to experience militarized disputes with each other. Just as importantly, these results are consistent whether or not we take into account the extent to which those states are directly dependent on each other in terms of trade. This means that the pacific effects of trade that result from joint membership in a trading community do not depend on the extent of direct trade dependence, which

Table 6.1 Logit models of MAOZMIDs

Variable	Low resolution		Med. resolution		High resolution	
	Model 1	Model 2	Model 3	Model 4	Model 5	Model 6
Same trading community	−0.327**	−0.336**	−0.299*	−0.301*	−0.442*	−0.465*
	(0.120)	(0.120)	(0.150)	(0.149)	(0.193)	(0.193)
Dyadic trade dependence low	−31.816*	–	−33.693*	–	−32.566	–
	(15.948)		(16.674)		(16.711)	
MaxFlow	−0.082	−0.128	−0.098	−0.149	−0.091	−0.140
	(0.395)	(0.399)	(0.402)	(0.407)	(0.406)	(0.411)
GDP high	0.241***	0.261***	0.257***	0.277***	0.253***	0.273***
	(0.052)	(0.051)	(0.053)	(0.052)	(0.052)	(0.051)
GDP low	0.157***	0.145***	0.154***	0.142***	0.153***	0.141***
	(0.039)	(0.039)	(0.040)	(0.039)	(0.040)	(0.039)
Democracy high	−0.004	−0.003	−0.007	−0.006	−0.006	−0.006
	(0.019)	(0.019)	(0.018)	(0.019)	(0.019)	(0.019)
Democracy low	−0.171***	−0.173***	−0.168***	−0.170***	−0.171***	−0.173***
	(0.025)	(0.026)	(0.025)	(0.026)	(0.026)	(0.026)
Shared IGO memberships	0.010	0.006	0.009	0.006	0.009	0.005
	(0.007)	(0.006)	(0.007)	(0.006)	(0.007)	(0.006)
Contiguity	1.734***	1.685***	1.764***	1.714***	1.787***	1.741***
	(0.203)	(0.205)	(0.203)	(0.206)	(0.206)	(0.209)
Distance (logged)	−0.659***	−0.645***	−0.649***	−0.634***	−0.645***	−0.632***
	(0.089)	(0.090)	(0.088)	(0.089)	(0.087)	(0.089)
Major power	1.532***	1.469***	1.505***	1.440***	1.517***	1.454***
	(0.185)	(0.181)	(0.185)	(0.182)	(0.185)	(0.182)
Alliance	0.378	0.313	0.405	0.341	0.432	0.375
	(0.222)	(0.228)	(0.227)	(0.234)	(0.228)	(0.235)
Capability ratio (logged)	−0.103**	−0.098*	−0.105**	−0.101*	−0.103*	−0.098*
	(0.040)	(0.040)	(0.040)	(0.040)	(0.040)	(0.040)
Peaceyears	−0.387***	−0.384***	−0.382***	−0.379***	−0.381***	−0.378***
	(0.044)	(0.044)	(0.044)	(0.044)	(0.044)	(0.044)
Constant	−3.946***	−4.076***	−4.253***	−4.399***	−4.257***	−4.383***
	(0.857)	(0.869)	(0.823)	(0.837)	(0.829)	(0.843)
N	383,126	383,126	383,126	383,126	383,126	383,126
χ^2	2,586.89***	2,636.76***	2,594.20***	2,640.70***	2,626.19***	2,667.13***

Robust standard errors in parentheses
* $p < 0.05$, ** $p < 0.01$, *** $p < 0.001$
Resolutions correspond to 3, 7 and 14 trading communities. Estimates for 3 temporal spline variables not reported

is the key explanatory variable in extant theories of trade of conflict. In terms of substantive effect, dyads in the same trading community are 48, 47, and 59 % less likely to experience a militarized dispute in Models 1, 3 and 5, respectively.

The control variables generally have the expected relationships with conflict and are consistent with the results of earlier studies [17, 42, 45]. Consistent with Oneal and Russett [42] and Gartzke [17], we find that DEMOCRACY LOW is associated with a lower probability of conflict, while DEMOCRACY HIGH is associated with a higher probability. Interestingly, unlike existing studies, we find that dyadic alliance relationships do not have a significant relationship with conflict. Taking only dyadic trade dependence into account, allies are less likely to fight wars, as Oneal and Russett [42] and Gartzke [17] found. However, when we also account for indirect trade dependence by including trading community membership in the model, allies are no less likely to fight than non-allies. This result suggests that trading community membership explains conflict behaviour to a sufficient extent as to obscure the effects of direct alliance links. In other words, it may be the case that indirect trade links drive the significant relationship between alliances and conflict found in other studies.

In addition to the results reported in Table 6.1, we used a model identical to Model 1 to test our hypothesis over other resolution levels. At all resolution levels between 3 and 15 trading communities, we find that SAME TRADING COMMUNITY has a significant ($p < 0.05$), negative relationship with MAOZMID, which indicates that, within a significant range, our results do not depend on the resolution level we specify. In other words, if we view the world as consisting of 3 very large trading communities, such as those defined in Fig. 6.3a, then joint membership in these communities is associated with a lower probability of conflict. Yet even if we take those communities and divide them into sub-communities, such as those defined in Fig. 6.3c (and even slightly smaller ones), joint membership in these smaller groups continues to be associated with a lower probability of conflict. We also follow Dorussen and Ward [14] in testing whether our results are consistent when we examine only "politically-relevant dyads", i.e., those that are either contiguous or include at least one major power. For this sample, we found that SAME TRADING COMMUNITY has a significant ($p < 0.05$) and negative relationship with MAOZMID for the same range of levels of aggregation as reported above (i.e., 3–15 trading communities).

Our results therefore demonstrate that across a broad range of plausible sizes of trading communities, dyads within these communities have a lower probability of conflict. Nonetheless, we find that joint membership in small groups is not significantly associated with a lower probability of conflict. In such small groups, there may not be sufficient (or any) members with the capacity to pay the costs of preventing a potential conflict. Similarly, we also find that joint membership in too large groups is not significantly associated with a lower probability of conflict. In groups so large and diverse, the group's members may not be capable of overcoming the collective action problem of preventing the conflict. These results therefore suggest that, while the relative density of trade links is an important predictor of conflict, this factor interacts with group size in ways that merit further investigation.

6.2 Scientific Communities and Negative Links

Although negative links are often present in networks, they are not always being discerned explicitly. For example, consider the internet, where web pages are linked through hyperlinks. Of course these links can be negative in its context, for example, "this guy [http://www.someguy.com] is a complete idiot". More often, insulting language is used in internet fora or in comments on news articles, and are often directed and even personal in nature. Assuming these links to be positive (or rather, ignoring completely they might be signed links) then renders understanding the network quite difficult. Nonetheless, this is a common assumption.

This assumption is not limited to online content. Science, for instance, is characterized by cooperation and benign disagreement, but sometimes also by epistemic rivalry. In democratic politics, disagreement with opponents is endemic as it is vital for political identity and to attract voters. Military alliances and conflicts, and economic collaboration and competition are examples already discussed. In social fields in general, people are embedded in a variety of cooperative and conflicting relationships, originating from, or leading to various groups.

As a case in point, Shwed and Bearman [54] recently used the modularity approach to study consensus formation in scientific communities. Not having scientific consensus on certain issues might prevent taking further action. For example, for some time there was a debate about whether smoking was cancerous or not, something considered rigorously proven nowadays. Knowing when there is scientific consensus on some subject and when not, might help understanding the difficulties in reaching scientific consensus. Although some disagreement will always be present in ordinary scientific practice, it should be distinguished from epistemic rivalries, where different paradigms may clash and there is little or no consensus at all.

Shwed and Bearman claimed their approach enables them to distinguish consensus and benign criticism on the one hand from epistemic rivalry on the other hand. Their data were scientific journal citations, which they interpreted for their modularity analysis as positive links. On the basis of these citation data, they determined scientific communities and their salience, i.e. the extent to which those communities stood out from a random network, as indicated by the raw modularity scores.

Let us go over some of the assumptions that Shwed and Bearman made to get their results. Their first assumption is that in "normal science" [33], most citations signal agreement. This assumption is entirely plausible [25] and there is significant support for it in the literature [9, 60]. Their second assumption is that the comparatively few citations that represent disagreement have no ramifications for the communities detected. We will demonstrate, in contrast, that a small proportion of negative citations can substantially perturb the results. Their third assumption is that epistemic rivalries between communities are marked by a lack of cross-community citations. In other words, contending factions largely ignore each other. They infer from this assumption that if the salience of communities diminishes, consensus increases. However, a lack of citations between groups does not necessarily imply opposing views. On the contrary, they might simply indicate that the communities have different interests, rather than having opposing views. Such groups detected in scientific citation networks are what we would call thematic communities, i.e. groups of scholars specializing in, and writing about the same themes or topics. They are less likely to be positional communities of scholars who agree with community members and disagree with other communities' views. This is a consequence of scholars citing mostly papers that they consider relevant, regardless of their (dis)agreement with those papers.

In the following, we will analyse these issues, and demonstrate that it is nearly impossible to analyse contention and consensus within or between communities when treating all links as positive, opposite to their suggestion. First, we analyse patterns of scientific citations on "smoking is cancerous" and on "solar radiation is cancerous", the latter being the same data that Shwed and Bearman used. As a baseline, we treat all links as positive, and test whether salient community differences arise when a small randomly chosen portion of the links is coded as negative. Second, we study the evolution of the "smoking is cancerous" field over time, by combining community detection with automated content analysis of the abstracts of the pertaining papers. This shows that these communities are indeed more likely to be thematic communities, i.e. scientific sub-disciplines that focus on different subjects, rather than positional communities that have different views. Third, we analyse a dataset on a public debate in politics wherein positive and negative links were distinguished during data coding. We show that the community structure obtained when—incorrectly—assuming all links to be positive is radically different from the community structure obtained when we properly distinguish between positive and negative links.

6.2.1 *Effect of Negative Links*

Let us now scrutinize Shwed and Bearman's assumption that the comparatively few citations that represent disagreement have no substantial impact on the communities detected. While some scientific citations are certainly critical, perhaps the proportion of negative references is so low that it is safe to assume that the comparatively few citations that represent disagreement have no impact on the communities found?

To see if this is the case we examine two cases: The "solar radiation is cancerous" and the "smoking is cancerous" datasets. We received the "solar radiation is cancerous" data from Uri Shwed, so these citations are exactly the same as they used in their paper. We collected the "smoking is cancerous" dataset from the ISI Web of Science using the same procedure Shwed and Bearman followed. For the latter data we also have the abstracts of most papers, allowing us to analyse to some extent the scientific content of the communities, which we can't for Shwed and Bearman's data.

To distinguish negative from positive references, we would have to acquaint ourselves with the vernacular of cancer researchers and read thousands of papers, which is beyond feasibility. So, to test the impact of negative links, we therefore set up the following simple procedure. We take a random sample of the links in the corpus, turn them into negative links, and perform community detection on that network. We repeat this procedure a hundred times for each year, and measure each time the difference between the "negatively modified" assignment of nodes into communities and the original assignment, and do this for 5 and 10 % of negative links, respectively. To quantify the similarity of the assignments, we use the measure of Normalized Mutual Information (NMI), as detailed in Sect. 2.4.2. In order to make sure that the

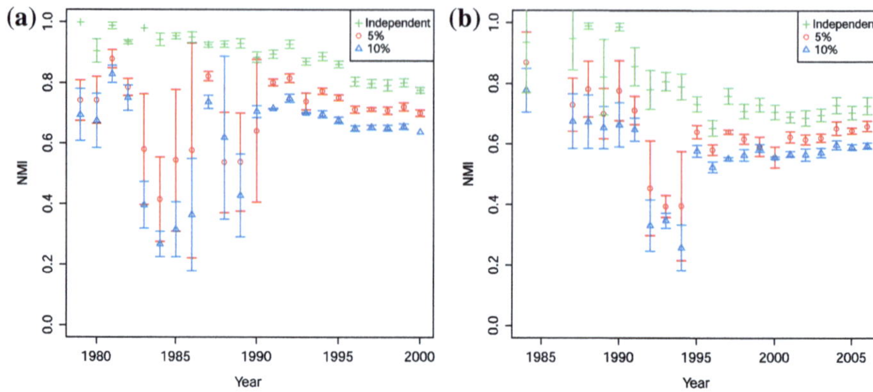

Fig. 6.4 Difference in communities when making links negative. **a** Smoking. **b** Solar radiation

observed differences do not arise because of the heuristic nature of the algorithm, which may lead to somewhat different outcomes in different runs, we perform the same comparison but without changing any of the links into negative.

The results are displayed in Fig. 6.4; vertical bars indicate variation and mean over 100 runs of the NMI score, and the comparison treatment with only positive links is called "independent." The figure shows that even a low proportion of negative links can cause assignments to differ more strongly than when all links are positive. This is the case both for the "smoking is cancerous" data Fig. 6.4a and for the "solar radiation is cancerous" data Fig. 6.4a. Obviously these differences become larger and more salient when the percentage of negative links increases. Our findings suggest that Shwed and Bearman's assumption that the comparatively few citations that represent disagreement have no impact on the communities detected is incorrect. We have shown that negative links do have an impact and cannot be ignored if one wants to study contention.

We may expect that in actuality, negative links will lead to even more pronounced differences. The reason is that by sampling a certain percentage of the links randomly, we ignored any pattern in the negative links, while we know from both social balance theory and empirical studies [57] that negative links tend to be present in between specific communities, not randomly throughout the network. Those networks with a small percentage of negative links are thus likely to have a more pronounced community structure than we find here. Moreover, such a small percentage of negative links is normally present in science. During periods of epistemic rivalry, when the percentage of negative links is higher, the difference will usually be larger. As said, the actual pattern of negative links is unknown to us and remains an empirical question. Nonetheless our analysis shows that a different community structure is likely to be detected when negative links are explicated.

6.2.2 *Dissensus or Specialization?*

Now let us focus on Shwed and Bearman's assumption that epistemic rivalry, i.e. a lack of consensus, is characterized by a lack of cross-community citations. To examine what these communities could represent with respect to a scientific field, we use the "smoking is cancerous" dataset which contains abstracts along with citations. We first extract all words used in all abstracts of the corpus. We assume that a group of articles that uses a shared vocabulary distinct from other groups discusses similar topics or methods. The common technique for extracting terms specific to a (set of) document(s) is the so-called Term Frequency-Inverse Document Frequency (tf-idf). Let w be some word (term), and let $n_w(d)$ be the number of occurrences of the word in some document d. Furthermore, let N_w represent the number of documents in which the word w occurs. Then the tf-idf is defined as

$$\text{tf-idf}(w, d) = \frac{n_w(d)}{\max_w n_w(d)} \log \frac{N}{N_w}$$

where N is the total number of documents. The underlying principle is that for a certain term to be of specific interest or salience in a document, it should be frequently mentioned in that specific document, and not that much elsewhere [34]. For the groups (that we first detect through modularity on the citation network), if terms are common in a specific group and rare elsewhere, this indicates that papers in that group concern similar themes or topics. At the group level, we focus on the five most salient terms according to this tf-idf measure. Moreover, by using the multi-slice modularity method [40], we obtain a dynamic view of the evolution of the groups, displayed in Fig. 6.5, together with the five most salient terms for each group. This graphical representation of group dynamics as an alluvial diagram was first used by [51]. To avoid a cluttered image, we only show the 12 largest groups, which over the period of observation have at least 1,000 papers. Within a group, it is possible that scholars criticize each other, but we can't detect contention since we do not know which citations are negative.

We can, however, see how the community structure changes over time with respect to common themes or topics. The different communities seem to focus on different research topics. Some communities seem to be researching different types of cancer, such as lung cancer, breast cancer, pancreatic cancer or colon cancer. Other communities seem to be (at least partly) founded on a common research background, such as the `p53` tumour suppressor [10] and the `gstm1` gene. Finally, some communities focus on two products of tobacco (smoke), namely two nitrosamines, `nnk` and `nnal` that are associated with risk of cancer. Most of the changes in communities seem to be due to switches and merges between related communities. Overall, they are relatively stable over time, and mostly seem to expand, pointing to an expansion of research, and possible intensification of scientific specialization.

This approach makes it possible for us to provide a more substantive description of the evolution of the community structure. Figure 6.5 provides evidence that the field self-organizes into thematic groups in a process of ongoing scientific specialization,

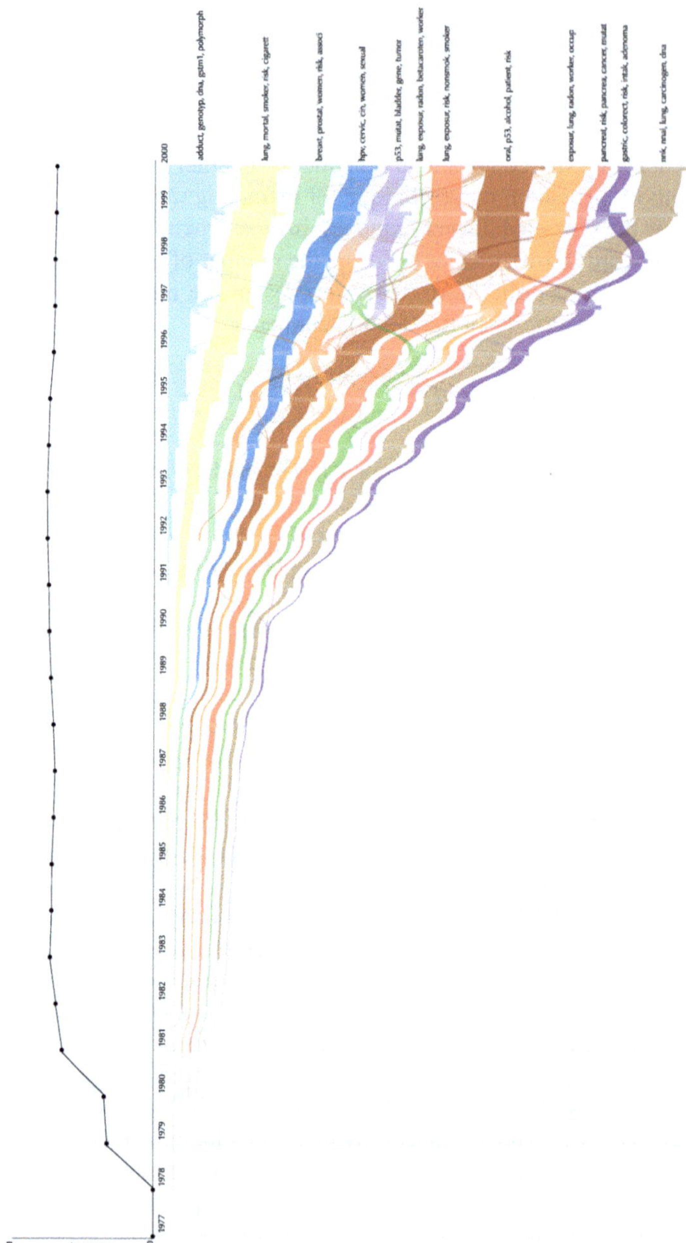

Fig. 6.5 Communities over time with most frequent terms

net of possible disagreements within these groups. This is a consequence of scientists citing papers they consider relevant, regardless of possible disagreements. It seems less likely that the groups detected are positional, consisting of scholars who mutually agree while disagreeing with other groups' views. In absence of information about negative links, not much can be said with any certainty about consensus or dissensus. It is possible that within thematic communities there is disagreement such that, once negative links are explicated, they turn out to be further partitioned into positional communities.

6.2.3 A Public Debate

We now show for data wherein we *can* distinguish positive and negative links how large the difference between community assignments can be when ignoring this distinction. Our dataset consists of references between authors in the debate about minority integration in the Netherlands. We focus on longer articles published in two broadsheet newspapers (*NRC Handelsblad* and *De Volkskrant*) in between the assassination of the populist politician Pim Fortuyn (6 May 2002) and the assassination of film maker Theo van Gogh (2 November 2004). We selected articles from the Lexis-Nexis database through the key word "integration" in conjunction with "foreigners", "Muslims" or "minorities". During this turbulent period in Dutch political history there were 149 long (over 1,000 words) articles on integration.

References to individuals (both Dutch and foreign, dead and alive) as well as to institutions (like political parties) or think tanks were manually coded by Uitermark [58]. In our 149 articles, the references were distinguished according to their tone: positive, neutral or negative. As a rule, positive and negative codes were assigned only if references were unambiguous. References were coded at the level of paragraphs, so it was possible for one article to contain several references to the same author, with each paragraph being coded according to the evaluation implicit or explicit in the reference(s).

In total 1,779 references by authors commenting on others were coded, either as positive (318), neutral (930), or as negative (531) directed links. Here we include only the positive and the negative links and we consider only the largest component of the network, which has 323 authors. The link weights between two authors were defined by subtracting the number of negative references from the number of positive references.

First, we identify communities while assuming that all references are positive. As a result, the network in Fig. 6.6 has a number of relatively dense groups of authors referring to each other.

When we distinguish positive and negative links in Fig. 6.7, consistent with the data, it turns out there are in fact two large communities, each with quite different membership than any of the communities from Fig. 6.6. There are many references between these two large communities but they are mostly negative: the two communities clearly disagree, and community membership now corresponds to a large

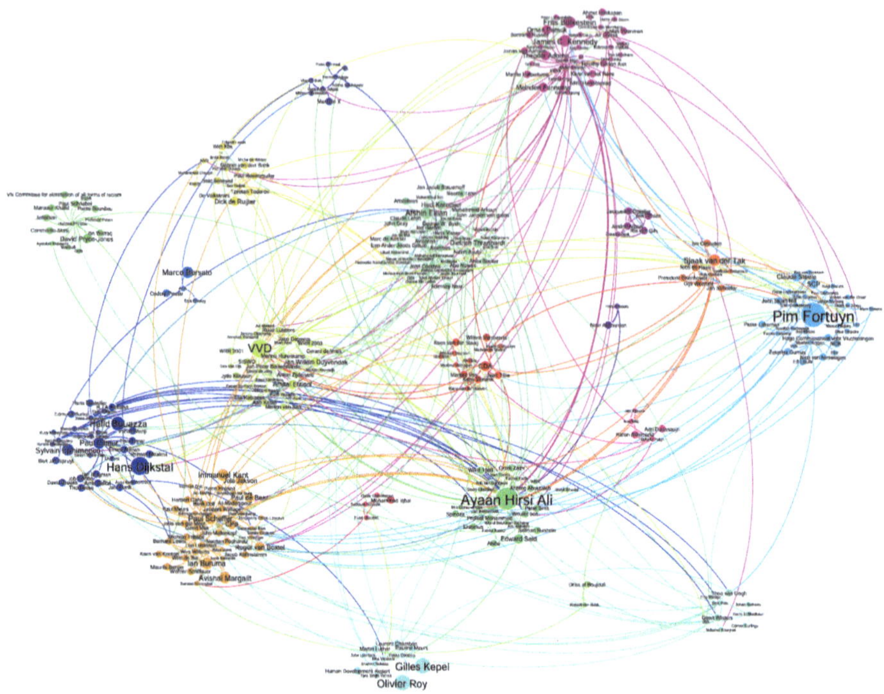

Fig. 6.6 Thematic communities assuming ties to be positive

extent to ideological identification. The communities are positional rather than thematic, and contention is a key mechanism of group formation in this field. The large community on the left consists mostly of authors who argue against the stigmatization of Islam and other minorities, while the large community on the right contains a majority of authors who argue that mass migration and (radical) Islam present a threat to Western civilization and to the Netherlands in particular.

The NMI score for Figs. 6.6 and 6.7 is 0.34, which is relatively low given that many positively connected authors who were together in Fig. 6.6 stay together in Fig. 6.7. Our key point here is that if one assumes all links to be positive, a very different result is obtained than if negative links are explicated. Once both positive and negative links are taken into account, then it becomes possible to analyse contention. Moreover the number of edges between the two largest communities is quite substantial (mostly negative), contrary to the assumption that there are few citations in between contending groups.

In conclusion, without distinguishing negative links explicitly, little can be said about the contentious community structure, for which further research is necessary. If anything, the data suggest that a key mechanism of group formation is specialization into sub-fields, while it seems less likely that the mechanism of contention, leading to rivalling camps, plays an important role.

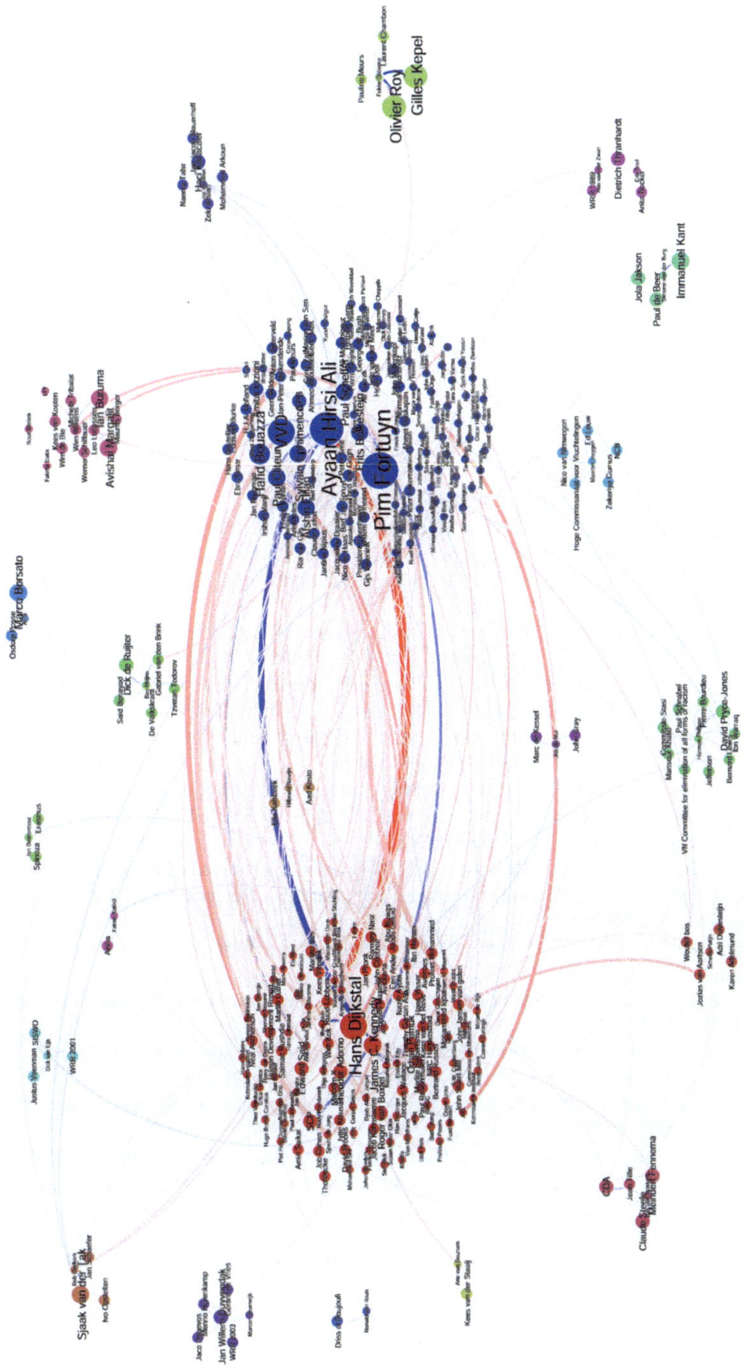

Fig. 6.7 Positional communities assuming ties to be negative

We showed that incorporating negative links in the analysis can have a substantial impact on the communities detected, even in fields where, as is the case for science, interaction is highly civilized and the proportion of negative references is low. So, in general, researchers have to explicate negative relations (criticism, repel, competition, or violence) when analysing fields wherein conflict is a mechanism of group formation. Only then can communities be detected validly, whatever method of detection is chosen.

References

1. Anderson JE (1979) A theoretical foundation for the gravity equation. Am Econ Rev 69(1): 106–116
2. Anderson JE, van Wincoop E (2003) Gravity with gravitas: a solution to the border puzzle. Am Econ Rev 93:170–192
3. Angell N (1933) The great illusion: a study of the relation of military power in nations to their economic and social advantage. W. Heinemann, London
4. Beck N, Katz JN, Tucker R (1998) Taking time seriously: time-series-cross-section analysis with a binary dependent variable. Am J Polit Sci 42(4):1260–1288
5. Böhmelt T (2009) International mediation and social networks: the importance of indirect ties. Int Interact 35(3):298–319
6. Böhmelt T (2010) The impact of trade on international mediation. J Conflict Resolut 54(4): 566–592
7. Bremer SA (1992) Dangerous dyads: conditions affecting the likelihood of interstate war, 1816–1965. J Conflict Res 36(2):309–341. doi:10.1177/0022002792036002005
8. Bremer SA (1993) Democracy and militarized interstate conflict, 1816–1965. Int Interact 18(3):231–249. doi:10.1080/03050629308434806
9. Case DO, Higgins GM (2000) How can we investigate citation behavior? A study of reasons for citing literature in communication. J Am Soc Inf Sci 51(7):635–645. doi:10.1002/(SICI)1097-4571(2000)51:7<635::AID-ASI6>3.0.CO;2-H
10. Cho Y, Gorina S, Jeffrey P, Pavletich N (1994) Crystal structure of a p53 tumor suppressor-DNA complex: understanding tumorigenic mutations. Science 265(5170):346–355. doi: 10.1126/science.8023157
11. Crescenzi MJC (2005) Economic interdependence and conflict in world politics. Lexington, Lanham
12. De Tocqueville A (2002) Democracy in America. Penguin Books Limited, London ISBN 9780141915692
13. Dorussen H, Ward H (2008) Intergovernmental organizations and the Kantian peace: a network perspective. J Conflict Res 52(2):189–212. doi:10.1177/0022002707313688
14. Dorussen H, Ward H (2010) Trade networks and the Kantian peace. J Peace Res 47(1):29–42. doi:10.1177/0022343309350011
15. Doyle MW (1986) Empires. Cornell University Press, Ithaca ISBN 9780801493348
16. Fukuyama F (1992) The end of history and the last man. Free Press (Macmillan), New York
17. Gartzke E (2007) The capitalist peace. Am J Polit Sci 51(1):166–191. doi:10.1111/j.1540-5907.2007.00244.x
18. Gasiorowski M, Polachek S (1982) Conflict and interdependence: east-west trade and linkages in the era of detente. J Conflict Res 26(4):709–729
19. Ghosn F, Palmer G, Bremer S (2004) The MID3 data set, 1993–2001: procedures, coding rules, and description. Conflict Manage Peace Sci 21:133–154
20. Gibler DM, Sarkees MR (2004) Measuring alliances: the correlates of war formal interstate alliance dataset, 1816–2000. J Peace Res 41(2):211–222. doi:10.1177/0022343304041061

21. Gilpin R (1981) War and change in World politics. Cambridge University Press, Cambridge ISBN 9780521273763
22. Gleditsch KS (2002) Expanded trade and Gdp data. J Conflict Res 46(5):712–724. doi: 10.1177/002200202236171
23. Gleditsch KS (2003) All international politics is local: the diffusion of conflict, integration, and democratization. University of Michigan Press, Ann Arbor ISBN 9780472023356
24. Gochman CS, Maoz Z (1984) Militarized interstate disputes, 1816–1976: procedure, patterns, and insights. J Conflict Res 28(4):585–615
25. Hanney S, Frame L, Grant J, BuxtonM, Young T et al (2005) Using categorizations of citations when assessing the outcomes from health research. Scientometrics 65(3):357–379
26. Henderson EA, Tucker R (2001) Clear and present strangers: the clash of civilizations and international conflict. Int Stud Q 45(2):317–338. doi:10.1111/0022-8833.00193
27. Hensel PR, Goertz G, Diehl PF (2008) The democratic peace and rivalries. J Polit 62(04):1173–1188. doi:10.1111/0022-3816.00052
28. Huntington SP (1996) The clash of civilizations and the remaking of world order. Simon & Schuster, New York ISBN 9781416561248
29. Jones DM, Bremer SA, Singer JD (1996) Militarized interstate disputes, 1816–1992: rationale, coding rules, and empirical patterns. Conflict Manage Peace Sci 15(2):163–213. doi:10.1177/073889429601500203
30. Kindleberger CP (1986) The World in depression, 1929–1939. University of California Press, Berkeley ISBN 9780520055926
31. Kissinger H (1994) Diplomacy. Simon & Schuster, New York ISBN 9781439126318
32. Krasner SD (1999) Sovereignty: organized hypocrisy. Princeton University Press, Princeton ISBN 9781400823260
33. Kuhn T (2012) The structure of scientific revolutions. The University of Chicago Press, Chicago ISBN 9780226458144
34. Manning CD, Raghavan P, Schütze H (2008) Introduction to information retrieval. Cambridge University Press, Cambridge ISBN 9780521865715
35. Maoz Z (2006) Network polarization, network interdependence, and international conflict, 1816–2002. J Peace Res 43(4):391–411. doi:10.1177/0022343306065720
36. Maoz Z (2009) The effects of strategic and economic interdependence on international conflict across levels of analysis. Am J Polit Sci 53(1):223–240. doi:10.1111/j.1540-5907.2008.00367.x
37. Maoz Z, Russett B (1992) Alliances, contiguity, distance, wealth, and political stability: is the lack of conflict among democracies a statistical artifact? Int Interact 17(3):245–268
38. Maoz Z, Russett B (2012) Normative and structural causes of democratic peace, 1946–1986. Am Polit Sci Rev 87(03):624–638. doi:10.2307/2938740
39. Marshall MG, Jaggers K (2002) Polity IV: regime authority characteristics and transitions datasets. http://www.systemicpeace.org/inscrdata.html
40. Mucha PJ, Richardson T, Macon K, Porter MA, Onnela JP (2010) Community structure in time-dependent, multiscale, and multiplex networks. Science 328(5980):876–8 (New York). doi:10.1126/science.1184819, arXiv:0911.1824
41. Olson M (1965) The logic of collective action: public goods and the theory of groups. Harvard University Press, Boston ISBN 9780674041660
42. Oneal J, Russett B (1997) The classical liberals were right: democracy, interdependence, and conflict, 1950–1985. Int Stud Q 41(2):267–293
43. Oneal J, Russett B (2005) Rule of three, let it be? When more really is better. Conflict Manage Peace Sci 22(4):293–310. doi:10.1080/07388940500339209
44. Oneal JR, Oneal FH, Maoz Z, Russett B (1996) The liberal peace: interdependence, democracy, and international conflict, 1950–85. J Peace Res 33(1):11–28. doi:10.1177/0022343396033001002
45. Oneal JR, Russett B (2011) The Kantian peace: the pacific benefits of democracy, interdependence, and international organizations, 1885–1992. World Polit 52(1):1–37. doi:10.1017/S0043887100020013

46. Oneal JR, Russett B, Berbaum ML (2003) Causes of peace: democracy, interdependence, and international organizations, 1885–1992. Int Stud Q 47(3):371–393. doi:10.1111/1468-2478.4703004
47. Pevehouse J, Nordstrom T, Warnke K (2004) The correlates of war 2 international governmental organizations data version 2.0. Conflict Manage Peace Sci 21(2):101–119. doi:10.1080/07388940490463933
48. Picciolo F, Squartini T, Ruzzenenti F, Basosi R, Garlaschelli D (2012) The role of distances in the World trade web. In: Proceedings of 2012 8th international conference on signal image technology and internet based systems. doi:10.1109/SITIS.2012.118, arXiv:1210.3269
49. Polachek SW (1980) Conflict and trade. J Conflict Res 24(1):55–78. doi:10.1177/002200278002400103
50. Polachek SW, Robst J, Chang YC (1999) Liberalism and interdependence: extending the trade-conflict model. J Peace Res 36(4):405–422. doi:10.1177/0022343399036004002
51. Rosvall M, Bergstrom CT (2010) Mapping change in large networks. PloS one 5(1):e8694. doi:10.1371/journal.pone.0008694, arXiv:0812.1242v1
52. Russett B, Oneal JR, Davis DR (2003) The third leg of the Kantian tripod for peace: international organizations and militarized disputes, 1950–85. Int Organ 52(03):441–467. doi:10.1162/002081898550626
53. Russett BM, Oneal JR, Cox M (2000) Clash of civilizations, or realism and liberalism deja vu? Some evidence. J Peace Res 37(5):583–608. doi:10.1177/0022343300037005003
54. Shwed U, Bearman PS (2010) The temporal structure of scientific consensus formation. Am Sociol Rev 75(6):817–840. doi:10.1177/0003122410388488
55. Singer JD, Small M (1966) Formal alliances, 1815–1939: a quantitative description. J Peace Res 3(1):1–31. doi:10.1177/002234336600300101
56. Small M, Singer JD (1969) Formal alliances, 1816–1965: an extension of the basic data. J Peace Res 6(3):257–282. doi:10.1177/002234336900600305
57. Szell M, Lambiotte R, Thurner S (2010) Multirelational organization of large-scale social networks in an online world. Proc Natl Acad Sci USA 107(31):13636–41. doi:10.1073/pnas.1004008107, arXiv:1003.5137
58. Uitermark J (2012) Dynamics of power in Dutch integration politics: from accommodation to confrontation. University of Amsterdam Press, Amsterdam ISBN 9789089644060
59. Ward MD, Gleditsch KS (1998) Democratizing for peace. Am Polit Sci Rev 92(1):51. doi:10.2307/2585928
60. White HD (2004) Citation analysis and discourse analysis revisited. Appl Linguist 25(1):89–116. doi:10.1093/applin/25.1.89

Part II
Social Balance and Reputation

Chapter 7
Social Balance

Negative links play a prominent role in many social scientific fields, although most research has almost exclusively focused on positive links. Ranging from stereotype formation [9, 16], norm maintenance [8] to social conflicts [12] and armed conflicts [14], in all situations negative links play a pivotal, if not primal, role. Often they constitute the first organizing principle in such networks, and sometimes the opposition between two contending groups is even stronger then their internal cohesion. However, how and why negative links form exactly is not completely clear. In this chapter we will investigate the structure of negative (and positive) links, and in the next chapter focus on their dynamics.

Although seemingly unrelated, negative links also play a natural role in the evolution of cooperation. We assume that if a link is negative, people do not cooperate, while if it is positive they do cooperate. Not everybody necessarily cooperates with everybody, and it is in fact often advantageous not to cooperate. At the same time, we do see a lot of cooperation in nature. This has long baffled social scientists and biologists alike, and remains an elusive problem. We will analyse the structure and dynamics of negative links in the context of the evolution of cooperation in Chap. 9.

But first we will turn to the concept of social balance [2–5, 10], which we already briefly saw while discussing negative links in community detection. We will discuss it in somewhat more detail than before. Social balance can be seen as the first organizing principle of networks with negative links. The basic idea is that triads (cycles of length 3) should be balanced: friends should think alike about a third person, while enemies should disagree. We will see that if all triads are balanced, a complete network can be split into two factions. Both [15] and [6] provide an introductory chapter on social balance.

The idea of social balance is often motivated from a theory known as cognitive dissonance, which is a theory in psychology that dictates that different beliefs and actions should be in accord with each other [7]. That is, if you think that saving wildlife is an important issue, yet you condone elephant hunting, this creates some friction, some dissonance. Of course, not all beliefs should be in perfect accord with each other, and some contradictory beliefs create more dissonance, while other create less dissonance.

V. Traag, *Algorithms and Dynamical Models for Communities and Reputation in Social Networks*, Springer Theses, DOI: 10.1007/978-3-319-06391-1_7,

The first step it to extend this idea to another person. Consider that you have some friend (or at least have that belief). Suppose that person, that friend, has very different ideas than you. It is then argued that this would induce some cognitive dissonance as well. After all, why should you be friends with somebody who is completely different from you? So, in order to reduce the amount of dissonance, two things can happen: either the friendship declines or beliefs of the two friends converge. So, in general we should expect most friends to think alike (at least to some extent).

The second step is now to consider a third person. If two friends are expected to think alike, the same should hold for their opinion concerning a third party. In particular, if somebody likes a third person, his friend is expected to also like that person, and similarly so if somebody dislikes that third person. A similar reasoning holds for two enemies. If somebody likes a third person, their enemy is expected to dislike that person.

We can thus discern between two types of triads: balanced ones and unbalanced ones, as illustrated in Fig. 7.1. Balanced triads are consistent and do not induce any stress in terms of dissonant beliefs or behaviour. Of course, if somebody is hated by a lot of people this will surely induce stress, but it will not be cognitively dissonant. In fact, you might wonder how somebody who is disliked that much thinks about himself, and indeed such a situation also often induces a negative self-esteem, congruent with cognitive dissonance theory [1, 11–13].

Unbalanced triads are believed to induce some stress due to cognitive dissonance, thereby creating an incentive for changing the unbalanced triad. For example, if Alice and Bob are good friends and Bob likes Eve, but Alice doesn't like her, this creates some tension. So, we might expect either Bob to change his relationship to either Eve or Alice, or expect Eve and Alice to become friends. In everyday life such situations might pop-up for example when a couple breaks up: people may have befriended both partners, but when the partners break up, their positive link is flipped to a negative link, thereby creating an unbalanced triad. This unbalanced situations is then often resolved by only staying friends with one of the two partners, or "choosing sides" so to speak.

Although the motivation comes from cognitive dissonance theory, the theory of social balance has been formalized to quite some extent without reference to cognitive dissonance. The focus is in first instance on triads and complete graphs that can be split into (at most) two factions. Similar definitions can also be provided on sparse graphs, and some weaker definition can be given such that a graph can be split into more factions.

7.1 Balanced Triads

The notion of social balance can be formalized by looking at triads (cycles of length 3) [10]. In general, we define a signed graph as $G = (V, E^-, E^+)$ where $E^- \subseteq V \times V$ are the negative links and $E^+ \subseteq V \times V$ the positive links and $E^- \cap E^+ = \emptyset$, so that no link is both positive and negative. Furthermore, we will restrict ourselves

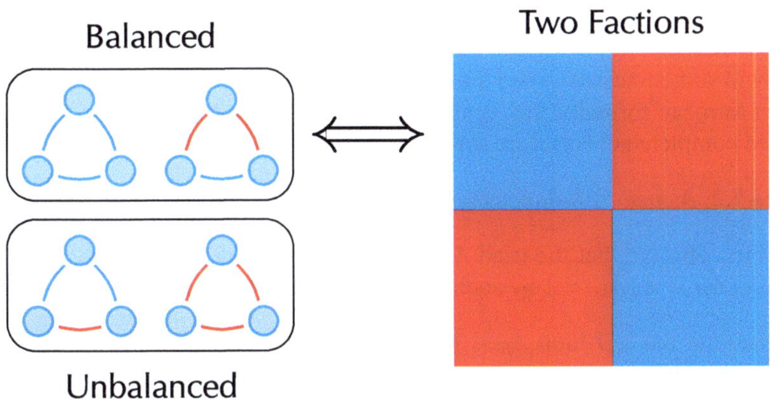

Fig. 7.1 Balanced triads in complete network

to undirected graphs. In this case, the definition of undirected is not immediately straightforward, but in this case we mean that if there is an edge $(i, j) \in E^+ \cup E^-$ then also $(j, i) \in E^+ \cup E^-$. So, the signs may in principle be different, but if there is a link in one direction, there must also be a link in the opposite direction. The adjacency matrices A^+ and A^- are then defined accordingly, so that $A^+_{ij} = 1$ whenever $(i, j) \in E^+$ and zero otherwise, and similarly $A^-_{ij} = 1$ if $(i, j) \in E^-$ and zero otherwise. The signed adjacency matrix will be denoted by $\tilde{A} = A^+ - A^-$. In this section we will work exclusively with the signed adjacency matrix, so that we will use $A = \tilde{A}$ in order to avoid cluttering of the notation. Hence, $\tilde{A} = A$ can be also defined as

$$A_{ij} = \begin{cases} -1 & \text{if } (i, j) \in E^-, \\ 1 & \text{if } (i, j) \in E^+, \\ 0 & \text{otherwise.} \end{cases} \tag{7.1}$$

Notice that in principle this matrix need not be symmetric, although it will follow that for socially balanced networks it is (and so undirected and sign symmetric is the same for socially balanced graphs). Nonetheless, because the network is undirected, we do have that $|A| = |A^\top|$. Social balanced graphs are then also sign symmetric so that we obtain $A = A^\top$. This can be relatively straightforwardly extended to weighted graphs, but for those we are still foremost concerned with their sign, which then simply reduces to the case under consideration.

In this section we will first concentrate on complete graphs (including self-loops), such that there is an edge for all pairs of nodes. That is $E^+ \cup E^- = V \times V$ and $|A_{ij}| = 1$ for all i, j. We will define a balanced triad as a cycle of three nodes, for which the product of the signs of the edges are positive.

Definition 7.1 A triad i, j, k is called (socially) balanced whenever

$$A_{ij}A_{jk}A_{ki} = 1. \tag{7.2}$$

A complete signed graph G is (socially) balanced whenever all triads are balanced.

We will sometimes also state that some real matrix X is socially balanced, which is taken to mean formally that $\text{sgn}(X)$ is socially balanced. Let us first prove that balanced complete networks are symmetric.

Lemma 7.2 *A socially balanced complete network is sign symmetric.*

Proof First observe that the triad $A_{ii}A_{ii}A_{ii} = 1$ is balanced, so that $A_{ii} = 1$. Then also $A_{ii}A_{ik}A_{ki} = A_{ik}A_{ki} = 1$ by social balance, so that $A_{ik} = A_{ki}$. □

Notice that the self-loop A_{ii} is pivotal here. This could be interpreted as self-esteem (what does i think of i?), and is always positive in a balanced network. If self-loops are not included, the signs might be of opposing sign if we only consider triads. However, in the next section we will see that the more general definition will preclude this case.

If a complete network is socially balanced, we can split such a graph in (at most) two factions, such that the links between the two factions are negative and the links within a faction are positive.

Definition 7.3 Let $G = (V, E^+, E^-)$ be a signed graph, then a faction $F \subset V$ is a subset of nodes such that

$$\begin{aligned}(u, v) &\notin E^- \text{ for } u, v \in F,\\ (u, v) &\notin E^+ \text{ for } u \in F, v \in V \setminus F.\end{aligned}$$

A partition into factions is then a set factions $\mathcal{F} = \{F_1, F_2, \ldots, F_q\}$ such that $V = \bigcup_{i=1}^{q} F_i$ and $F_i \cup F_j = \emptyset$ for $i \neq j$, similar to the partition into communities (see p. 13). Notice that a partition into factions corresponds to a block partition of a matrix (up to reordering) as

$$A = \begin{pmatrix} + & - & \cdots & - \\ - & + & \cdots & - \\ \vdots & \cdots & \ddots & \vdots \\ - & \cdots & - & + \end{pmatrix}$$

where $+$ denotes a block of only non-negative entries (i.e. 0 or 1) and $-$ denotes a block of non-positive entries (i.e. 0 or -1). For a socially balanced signed graph, this partition is limited to at most two factions (this restriction is not present for weak social balance, see Sect. 7.3). This condition can also be expressed in terms of the spectrum. This idea is illustrated in Fig. 7.1.

Theorem 7.4 *Let $G = (V, E^+E,^-)$ be a complete signed graph with symmetric adjacency matrix A. Then the following are equivalent:*

1. *G is socially balanced,*
2. *G can be split in at most two factions,*
3. $A = uu^\top$ *where* $|u_i| = 1$.
4. $\lambda_1(A) = n$, $\lambda_i(A) = 0$ *for* $i \geq 2$.

Proof Let us first prove that (1) $\Rightarrow$ (2). Let us take a node $v \in V$, and set $F_1 = v \cup N^+(v)$ where $N^+(v) = \{u|(u,v) \in E^+\}$ are the positive neighbours of v, and set $F_2 = V \backslash F_1$. Suppose $(u,v) \in E^-$. Then if u and v are in the same component they have a common positive neighbour w, and hence the triangle uvw has negative sign, contradicting social balance. Hence all negative links are between F_1 and F_2 and all positive links within F_1 and F_2. Indeed, this corresponds to two factions (one which might be empty). Now let us prove (2) $\Rightarrow$ (3). Notice that up to relabelling the split in at most two factions corresponds to

$$A = \begin{pmatrix} + & - \\ - & + \end{pmatrix} = (+ \mid -) \begin{pmatrix} + \\ - \end{pmatrix}. \tag{7.3}$$

where each $+$ and $-$ corresponds to a block of only 1 and -1 respectively. So with $u = (+ \mid -)$ this gives the requested property. Then (3) $\Rightarrow$ (4) is almost immediate. Since $A = uu^\top$ we obtain that $Au = uu^\top u = un$ so that $\lambda_1(A) = n$. Since A is a rank one matrix $\lambda_i(A) = 0$ for $i \geq 2$. Finally, let us prove (4) $\Rightarrow$ (1). Since A is a rank one matrix we can write $A = uu^\top$ for some u, so that $A_{ij} = u_i u_j$ and $u_i^2 = 1$. Then $A_{ij}A_{jk}A_{ki} = u_i u_j \, u_j u_k \, u_k u_i = 1$. □

Notice that if $A = uu^\top$ then $A^2 = nuu^\top$ so that all powers of A are also socially balanced (or strictly speaking $\text{sgn}(A^2)$ is balanced).

This theorem completely describes socially balanced complete signed graphs, and their structure is very simple. Notice that the eigenvalue decomposition of $A = uu^\top$ corresponds exactly to the minimization of the frustration provided in Sect. 5.1.1 on p. 94.

7.2 Balanced Cycles

If we analyse signed graphs that are not complete, the definition of balanced triads is no longer satisfactory. After all, if a link is missing, such that $A_{ij} = 0$, is that triad involving the link (ij) then not balanced? Suppose we say we only take into account triads which are complete (all three links are present). Then there are plenty examples of signed graphs such that each triad is balanced, but that it no longer neatly splits into two factions. So, for general signed graphs a somewhat different definition of social balance is appropriate [10]. However, it is consistent with the case of complete signed graphs in the previous section, and it emerges as a special case.

The focus in this case is on cycles, and we can define a balanced cycle as follows.

Definition 7.5 Let $G = (V, E^+, E^-)$ be a signed graph and A the signed adjacency matrix. Let $C = v_1 v_2 \dots v_k v_1$ be a cycle consisting of nodes v_i with $v_{k+1} = v_1$. Then the cycle C is called balanced whenever

$$\text{sgn}(C) := \prod_{i=1}^{k} A_{v_i v_{i+1}} = 1. \tag{7.4}$$

A signed graph G is called balanced if all its cycles C are balanced.

We will also call balanced cycles positive cycles, and unbalanced cycles negative cycles. Furthermore, we can define the sign of a path.

Definition 7.6 Let $P = v_1 v_2 \dots v_k$ be a path in a signed graph G with signed adjacency matrix A. The sign of the path P is then defined as

$$\text{sgn}(P) := \prod_{i=1}^{k-1} A_{v_i v_{i+1}}. \tag{7.5}$$

We then speak of a positive or negative path. Clearly a positive cycle consists of two paths of the same sign. The interpretation of balanced cycles remains similar as before. Consider for example a cycle of length 4 with a single negative link between node u and v. Then on the one hand there is a complete positive path between u and v (of length 3), while on the other hand there is a negative link between u and v. If we extend the previous idea that a friend of a friend is a friend to the third degree, u and v should be friends, not enemies. Hence, such a cycle should be unbalanced, and indeed its sign is -1. Now consider again a cycle of length 4 between u and v but now with one path consisting of two positive links and the other one of two negative links. Then u is the friend of a friend on the one hand, while it is the enemy of an enemy on the other hand. This is consistent with each other, and so the cycle is balanced.

It is immediately clear that a balanced undirected network should be sign-symmetric, since also the cycles of length 2 must be positive. Moreover, if there are any self-loops they need to be positive (them being cycles of length 1), similar to the previous section.

Lemma 7.7 *A balanced signed graph G is sign symmetric $A = A^\top$ and any self-loop is positive.*

From here onwards a socially balanced network is hence sign symmetric. Notice that this is consistent with the triads on a complete graph, but that we now also look to cycles of a different length. Notice there can be many cycles (particularly for complete graphs). However, it is possible to prove that if all chordless cycles (of size at least 3) are balanced, then all cycles are balanced. A chord is an edge between two vertices of a cycle, see also Fig. 7.2.

Fig. 7.2 Cycle with chord

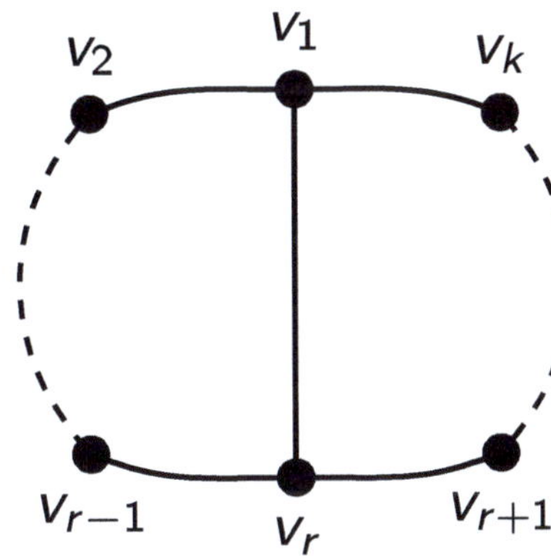

Lemma 7.8 *Let $C = v_1 v_2 \dots v_k v_1$ be a cycle with a chord between nodes v_1 and v_r in C (which by the previous lemma is sign symmetric). Then let $C_1 = v_1 v_2 \dots v_r v_1$ and $C_2 = v_1 v_{k-1} \dots v_r v_1$ be the induced subcycles. Then C is balanced if C_1 and C_2 are balanced*

Proof See Fig. 7.2 for an illustration. Let $s_1 = \text{sgn}(C_1)$ and $s_2 = \text{sgn}(C_2)$ be the sign of the cycles C_1 and C_2. By assumption we have that $s_1 = s_2 = 1$, since they are balanced. Then the path from v_1 to v_r in C_1 and the path from v_1 to v_r in C_2 must both have the same sign, since both cycle share the common edge (v_1, v_r). Since the signs of both paths from v_1 to v_r are the same, the cycle C must have positive sign. □

The inverse is not true however. A cycle may contain subcycles of negative sign. However, clearly, if all chordless cycles (of size at least 3) have positive sign, all cycles have positive sign, and so the network is balanced. Moreover, if a network is complete, the only chordless cycles are triads. Hence, a complete network is socially balanced if and only if all triads are balanced, so that indeed this more general definition is consistent with our discussion in the previous section.

Using this definition, we can prove that a balanced signed graph can be split into (at most) two factions. However, the eigenvalues and eigenvector might be more complicated because of the zeros in the signed adjacency matrix. Nonetheless, the largest eigenvector provides the information on the two factions (as was also shown in Sect. 5.1.1 on p. 94).

Theorem 7.9 *Let $G = (V, E^+, E^-)$ be a connected signed graph and A the signed adjacency matrix. Then G is balanced if and only if G can be split into (at most) two factions.*

Proof First assume G is balanced. Then pick a random $v \in V$ and set $F_1 = \{u|\text{sgn}(u - v \text{ path}) = 1\}$, i.e. all the nodes that can be reached through a positive path. Define $F_2 = V \backslash F_1$. Let $e = (u, w) \in E^-$. By construction $E^-(F_1) = \emptyset$ since it would otherwise contradict balance. Suppose now that $e \in E^-(F_2)$. Both the $u - v$ path and the $w - v$ path is negative (otherwise u and w would be in F_1). There is then a $u - w$ path that is positive (the product of the two negative paths is positive), and $(u, w) \in E^-$ so that there is a negative cycle, which contradicts balance. Hence, all negative edges lie between F_1 and F_2 so that any positive edges lie within F_1 and F_2, corresponding to two factions (one of which may be empty). Vice-versa, suppose G can be decomposed into two factions F_1 and F_2. Let C be a cycle. If C is contained within F_1 or F_2 it is completely positive, so that $\text{sgn}(C) = 1$. So, suppose C has some node $u \in F_1$ and $v \in F_2$. Then by definition of factions, both $u - v$ paths are negative, and so the cycle C is positive. Hence, all cycles are balanced, and so G is balanced. □

Whether social balance holds can be easily seen from the dominant eigenvector.

Theorem 7.10 *Let G be a connected signed graph and let u be the dominant eigenvector of the signed adjacency matrix A. Then G is balanced if and only if $F_1 = \{i \in V | u_i > 0\}$ and $F_2 = V \setminus F_1$ defines the split into two partitions.*

Proof If the split defines correct factions, then obviously G is balanced by the previous theorem. The other way around, suppose G is balanced. Then $A = A^T$. Let u be the dominant eigenvector. Suppose that $u_i A_{ij} u_j < 0$ for some i, j. Then let x be another vector with $|x_i| = |u_i|$ for all i and $x_i A_{ij} x_j > 0$ for all i, j, which is possible by social balance of G. Then $\|x\| = \|u\|$ and

$$u^\top Au = \sum_{ij} u_i A_{ij} u_j \tag{7.6}$$

$$< \sum_{ij} |u_i A_{ij} u_j| \tag{7.7}$$

$$= \sum_{ij} |x_i A_{ij} x_j| \tag{7.8}$$

$$= \sum_{ij} x_i A_{ij} x_j = x^\top Ax, \tag{7.9}$$

which contradicts the fact that u is the dominant eigenvector. Hence, we obtain that $u_i A_{ij} u_j \geq 0$ for all i, j and it defines a correct partition. □

Furthermore, if a signed adjacency matrix A is socially balanced, then so is A^2 (and by extension A^k for any k).

Theorem 7.11 *Let A be a balanced signed adjacency matrix. Then A^2 is balanced.*

Proof Let A be balanced. Suppose that i and j are in the same faction. Then all paths between i and j are positive, so that in particular $\sum_k A_{ik} A_{kj} \geq 0$. Now suppose that i and j are in a different faction, then all paths between i and j are negative, so that $\sum_k A_{ik} A_{kj} \leq 0$. Hence, A^2 is also balanced. □

Notice that the inverse does not hold, since the number of balanced cycles may outweigh the number of unbalanced cycles, so that A^2 might be balanced, while A is not. Consider for example a graph with the following sign structure

$$A = \begin{pmatrix} - & + \\ + & - \end{pmatrix}.$$

Then A^2 is indeed balanced, while A is not. Finally, it is obvious that any subgraph of a balanced graph is balanced itself.

7.3 Weak Social Balance

In the previous sections, any cycle with an odd number of negative links is called unbalanced. However, if you consider to split such a cycle into factions, there is not necessarily a problem. Consider for example the simplest case, a triad with three negative links. Although this triad is clearly not balanced (the cycle is negative), it can be easily split into three factions: each node constitutes its own faction. This is obviously a correct partitioning: there are only negative links between factions and positive links within factions (none in this case). Is it possible to characterize networks that can be split into factions in a simple way? The answer is yes, and cycles are still key [3, 5].

Consider as an example a cycle with a single negative link between node u and v. On the one hand node u and v should be clustered in the same faction since they are joined by an all positive path. On the other hand, node u and v should be clustered in a different faction, since they are joined by a negative path. This simple condition is in fact sufficient to characterize "clusterable" signed graphs.

Definition 7.12 A cycle $C = v_1v_2 \dots v_kv_1$ is termed weakly balanced if it contains not exactly a single negative link. A signed graph G is called weakly balanced if all its cycles C are weakly balanced.

We will also refer to the social balance defined previously as "strongly balanced". Any graph that is strongly balanced is then obviously also weakly balanced. A cycle that contains an even number of edges (has a positive sign) surely won't have exactly a single negative link. The inverse is not true however, the triad with three negative link being a prime counter example. This explains also why the two different definition might be called strongly and weakly balanced: strong balance is more constrained and implies weak balance, but not vice-versa. Notice that weakly balanced graphs must also be sign-symmetric (the cycle of length 2 must not have exactly one negative link).

Similar as before, we can focus on chordless cycles.

Lemma 7.13 *Let $C = v_1v_2 \dots v_kv_1$ be a cycle with a chord between nodes v_1 and v_r in C. Then let $C_1 = v_1v_2 \dots v_rv_1$ and $C_2 = v_1v_{k-1} \dots v_rv_1$ be the induced subcycles. Then C is weakly balanced if C_1 and C_2 are weakly balanced.*

Proof Both C_1 and C_2 do not contain a single negative link. Suppose that the link $v_1 v_r$ is not a negative link, then the only negative links of C_1 and C_2 also appear in C so that C is weakly balanced. Suppose that $v_1 v_r$ is a negative link. Then both C_1 and C_2 should contain at least one negative link that also appears in C so that C again is weakly balanced. □

The inverse is not true, which can be readily seen by considering an all-positive cycle with a single negative chord. The all-positive cycle clearly is weakly balanced, but the induced subcycles are not.

Theorem 7.14 *Let G be a signed graph. Then G is weakly balanced if and only if it can be split into factions* $\mathcal{F} = \{F_1, F_2, \ldots, F_q\}$.

Proof Suppose G is weakly balanced. Let $G^+ = (V, E^+)$ be the positive part of the signed graph, and let the factions be defined by the connected components of G^+. Suppose that $(u, v) \in E^-$. Then u and v are never in the same positive component since there would then be an all positive path between u and v, contradicting weak balance. Hence, this corresponds to a correct split into factions. Vice versa, suppose G is split into factions. Then any cycle clearly cannot contain exactly a single negative link, since there would then be a negative link within a faction, contradicting its definition. Hence G is weakly balanced. □

For a complete signed graph, the condition for it to be split into factions is even simpler.

Corollary 7.15 *Let G be a complete signed graph. Then G can be split into factions if and only if the triad composed of a single negative link is not present.*

Proof By lemma 7.13 we can look only to triads, and by the previous theorem these triads should not contain a single negative link. □

So, for a complete signed graph, there is only a single forbidden triangle for it to be weakly balanced: the triad with a single negative link. If in addition the triad with three negative links is a forbidden subgraph, it is strongly balanced.

The characterization of a signed graph that can be partitioned into factions is rather simple: the forbidden subgraphs are cycles with a single negative link. On the contrary, determining the minimum number of factions necessary is less trivial. To see this, let us constructs a contracted graph $G^* = (V, E)$ as follows. For every positive connected component (i.e. the components of G^+) we define a node $v \in V(G^*)$. Whenever there is a negative link between two connected components in G represented by two nodes $u, v \in V(G^*)$ we create a link $(u, v) \in E(G^*)$. Obviously, the graph G is weakly balanced, if and only there is no self-loop in G^*, corresponding to a cycle with a single negative link. After all, such a self-loop would imply that two nodes in a positive component would have a negative link, so that there must be a cycle with a single negative link.

Definition 7.16 Let G be a weakly balanced signed graph, so that it can be partitioned into factions $\mathcal{F} = \{F_1, \ldots, F_q\}$. We denote by $\chi(G)$ the minimum number of factions $q = |\mathcal{F}|$ necessary to partition the signed graph G into factions, and by $\Omega(G)$ the maximum number of factions, or

$$\chi(G) = \min |\mathcal{F}|$$
$$\Omega(G) = \max |\mathcal{F}|$$

The correspondence with the chromatic number $\chi(G)$ (the minimum number of colours necessary to colour a graph) is intentional, and there is a natural correspondence between the two. The upper bound $\Omega(G)$ is easily determined, and is provided by the number of nodes in G^*, i.e. by the number of positive components of G. Although for G^* this is the same as usual for any valid colouring, but for G the maximal number of colours is more restricted than usual in a colouring. The lower bound $\chi(G)$ can also be simply expressed by G^* and we have that $\chi(G) = \chi(G^*)$. So, the minimum number of factions necessary to partition a balanced signed graph G is equal to the chromatic number of the contracted graph G^*. The correspondence between a colouring problem and the partition into factions is not exact however. A graph can always be coloured, and always possesses a correct colouring. A signed graph however can not always be coloured, that is, partitioned into factions, since only weakly balanced graphs can be correctly partitioned.

Although the problem of determining the minimum number of factions reduces to determining the chromatic number $\chi(G^*)$ of the contracted graph G^*, this is still not easily determined or characterized. Two well known bounds exist and $\omega(G^*) \leq \chi(G^*) \leq \Delta(G^*)+1$ where $\omega(G^*)$ is the largest clique in G^* and $\Delta(G^*)$ the maximum degree. In general, it is NP-hard to determine $\chi(G^*)$, and no simple characterization is known to exist.

An easy case however is determining if $\chi(G^*) = 2$ in which case G^* is simple bipartite. There is a simple characterization of bipartite graphs, namely that they do not possess any odd-length cycles. This corresponds exactly to not having cycles of negative sign in the signed graph G (an odd number of negative links), and so corresponds to strong balance.

For a strongly balanced signed adjacency matrix A, we know that A^2 is also balanced. This is mainly due to the fact that products of any cycles are positive, so that any paths have signs consistent with social balance. For weak balance this no longer holds. For sparse graphs this is immediately clear, since not all links need to change in the same way when taking A^2.

Even for full graphs this does not hold however, which might not be readily clear. As a small counter example, consider three factions of size n_i, n_j and n_k (there might be other factions still) in a weakly balanced complete graph of size n. Then A^2 will have negative links between factions i and j yet have positive links between factions i and k and j and k, and so will no longer be socially balanced (neither weakly nor strongly), under the following conditions:

$$n_i + n_j > \frac{n}{2},$$
$$n_i + n_k < \frac{n}{2},$$
$$n_j + n_k < \frac{n}{2}.$$

In terms of common enemies, these inequalities are quite intuitive. It implies that both factions i and k and factions j and k have relatively many common enemies (i.e. $n_i + n_k < n/2$ and $n_j + n_k < n/2$), yet factions i and j don't have that many common enemies (i.e. $n_i + n_j > n/2$). This results in an A^2 with (ik) and (jk) being positive links, whereas (ij) is a negative link. By corollary 7.15 it is not weakly balanced (nor is it strongly balanced).

Summarizing, we have the following. Strongly balanced graphs

- have cycles of positive sign, and
- can be split into (at most) two factions,

whereas weakly balanced graphs,

- have cycles with not exactly one negative link, and
- can be split into factions.

Furthermore, social balance implies weak balance, but not vice-versa.

References

1. Bearman PS, Moody J (2004) Suicide and friendships among American adolescents. Am J Public Health 94(1):89–95. doi:10.2105/AJPH.94.1.89
2. Cartwright D, Harary F (1956) Structural balance: a generalization of Heider's theory. Psychol Rev 63(5):277–293. doi:10.1037/h0046049
3. Cartwright D, Harary F (1968) On the coloring of signed graphs. Elem Math 23(4):85–89. doi:10.5169/seals-26032
4. Cartwright D, Harary F (1979) Balance and clusterability: an overview. Academic Press, New York
5. Davis JA (1967) Clustering and structural balance in graphs. Hum Relat 20(2):181–187. doi: 10.1177/001872676702000206
6. Easley D, Kleinberg J (2010) Networks, crowds, and markets: reasoning about a highly connected world. Cambridge University Press, Cambridge
7. Festinger L (1957) A theory of cognitive dissonance. Stanford University Press, Stanford 9780804709118
8. Friedkin NE (2001) Norm formation in social influence networks. Soc Netw 23(3):167–189. doi:10.1016/S0378-8733(01)00036-3
9. Greenwald AG, Banaji MR, Rudman LA, Farnham SD, Nosek BA et al (2002) A unified theory of implicit attitudes, stereotypes, self-esteem, and self-concept. Psychol Rev 109(1):3–25. doi:10.1037/0033-295X.109.1.3
10. Harary F (1953) On the notion of balance of a signed graph. Mich Math J 2(2):143–146. doi: 10.1307/mmj/1028989917

11. Labianca G, Brass DJ (2006) Exploring the social ledger: negative relationships and negative asymmetry in social networks in organizations. Acad Manage Rev 31(3):596–614. doi: 10.5465/AMR.2006.21318920
12. Labianca G, Brass DJ, Gray B (1998) Social networks and perceptions of intergroup conflict: the role of negative relationships and third parties. Acad Manage J 41(1):55–67. doi:10.2307/256897
13. Lakey B, Tardiff TA, Drew JB (1994) Negative social interactions: assessment and relations to social support, cognition, and psychological distress. J Soc Clin Psychol 13(1):42–62. doi: 10.1521/jscp.1994.13.1.42
14. Maoz Z (2006) Network polarization, network interdependence, and international conflict, 1816–2002. J Peace Res 43(4):391–411. doi:10.1177/0022343306065720
15. Wasserman S, Faust K (1994) Social network analysis. Cambridge University Press, Cambridge
16. Wert SR, Salovey P (2004) A social comparison account of gossip. Rev Gen Psychol 8(2):122–137. doi:10.1037/1089-2680.8.2.122

Chapter 8
Models of Social Balance

There are two big questions associated with social balance. The first is how to detect whether a network can be split into factions. We already addressed this in Chap. 5. The second question concerns the emergence (and stability) of social balance. If indeed social balance should emerge from some process such as cognitive dissonance, can we model such a process and show that social balance results? In other words, what model can potentially lead to social balance? And under what conditions will it lead to social balance? In this chapter we will address these questions.

We will restrict ourselves to complete graphs and strong social balance. Of course, it would be very interesting to study models on sparse graphs and relate them to weak social balance. There do seem to be possibilities for addressing both issues in the modelling, yet the analysis becomes much more difficult, and is reserved for future work.

8.1 Discrete Models

One of the first suggested models for social balance was discrete in nature [2]. All links simply have a sign $+$ or $-$ (or $+1$ and -1), and the signs can only flip from positive to negative, or the other way around. Starting from some initial condition, the idea is then that if we flip edges long enough social balance may emerge. Of course, if we simply randomly flip links, this is unlikely to happen. So, these flips should follow some rules, which should obviously be related to social balance.

As stated, we will limit ourselves to complete graphs (without self-loops), so that we can focus on triads only. We will also assume that the graphs are undirected and sign symmetric, so that there are simply two balanced triads (those with 0 or 2 negative links), and two unbalanced triads (those with 1 or 3 negative links). Indeed, by Lemma 7.8 we should only concern ourselves with triads, and it is not necessary to focus on other cycles. This makes it significantly easier to analyse models for complete graphs than for general graphs. Moreover, we will only analyse the mean-field behaviour, which was shown to accurately predict simulated results [2].

V. Traag, *Algorithms and Dynamical Models for Communities and Reputation in Social Networks*, Springer Theses, DOI: 10.1007/978-3-319-06391-1_8,

Now the model goes as follows. We randomly choose a triad and if the triad is unbalanced, we will change one of its links so that is becomes balanced. However, it might be that some other triads then become unbalanced. So it is not immediately clear whether such a process will lead to social balance. There are then two choices available: we simple change one of the links randomly without concern for the other triads, known as Local Triad Dynamics (LTD), and one in which we only update if it improves the overall number of balanced triads, known as Constrained Triad Dynamics (CTD).

8.1.1 Local Triad Dynamics

Let us first detail the Local Triad Dynamics (LTD) model. Let us denote by Δ_k a triad with k negative links, so that Δ_0 and Δ_2 are balanced, and Δ_1 and Δ_3 are unbalanced. Whenever we choose a random triad, which is unbalanced, we flip one of its links, such that with probability p a triad changes from $\Delta_1 \to \Delta_0$, and with probability $1-p$ a triad changes from $\Delta_1 \to \Delta_2$. Obviously, with a single flip a triad Δ_3 can only change to Δ_2, since it has 3 negative links. We will first analyse how this model behaves in the limit of large n number of nodes.

Let us denote by $T = \binom{n}{3}$ the number of triads in a network, and by T_k the number of triads with k negative links, and define the proportion of triads with k negative links $t_k = T_K/T$. Furthermore, denote by m^+ and m^- the number of positive and negative links. Then obviously,

$$m^+ = \frac{1}{n-2}\sum_k (3-k)T_k \qquad m^- = \frac{1}{n-2}\sum_k kT_k,$$

since each Δ_k triad contains k negative links, and each link appears in $n-2$ triads. The density of positive links is then $\rho = m^+/m = \sum_k (3-k)t_k/3$, and obviously the density of negative links is $1-\rho = m^-/m = \sum_k kt_k/3$.

Finally, let us write by T_k^+ the average number of positive links that are attached to a Δ_k triad, which can be written as $T_k^+ = (3-k)T_k/m^+$, because each Δ_k triad has $3-k$ positive links, and there are m^+ positive links in total. The probability that a positive link is attached to a Δ_k triad is then

$$t_k^+ = \frac{T_k^+}{N-2} = \frac{(3-k)t_k}{\sum_l (3-l)lt_l}.$$

Similarly we can define T_k^- as the average number of positive links attached to a Δ_k triad, which is $T_k^- = kT_k/m^-$ and

$$t_k^- = \frac{T_k^-}{N-2} = \frac{kt_k}{\sum_l lt_l}$$

denotes again the probability a negative link is attached to a triad with k triads.

Suppose we have chosen an unbalanced triad Δ_1 with a single negative link at random. The probability of selecting such a triad is t_1. The probability that a negative link changes to a positive one is p, since $\Delta_1 \to \Delta_0$ with probability p, while the probability that a positive link changes to a negative one is $1 - p$. Suppose that $\Delta_1 \to \Delta_0$ and that a negative link switched sign. The probability that a negative link is attached to a Δ_k triad is t_k^- and these triads also change. So, in this case, the proportion of triads of type Δ_1 changes as

$$t_1' = t_1 - t_1^- + t_2^- - 1/T.$$

Since $1/T \to 0$ for large n, we ignore this contribution.

We can examine the other cases similarly. In general then, the probability a link flips from positive to negative can be calculated as $\pi^+ = (1-p)t_1$ since the probability to select a Δ_1 triad is t_1, and the flip $\Delta_1 \to \Delta_0$ happens with probability $1-p$. The probability a link flips from negative to positive is then $\pi^- = pt_1 + t_3$ since $\Delta_1 \to \Delta_0$ happens with probability p, and if we have selected a triad Δ_3 (which happens with probability t_3), it will always switch from negative to positive (as Δ_3 does not have positive links). We approximate the original dynamics by a continuous time differential equation.

Let us first look how the proportion of Δ_0 changes. With probability π^- a link changes from negative to positive. The probability a negative link is attached to a triad of type Δ_1 is t_1^-, so the probability that $\Delta_1 \to \Delta_0$ is $\pi^- t_1^-$. With probability π^+ a link changes from positive to negative, and the probability a positive link is attached to a triad of type Δ_0 is t_0^+. So with probability $\pi^+ t_0^+$ a triad $\Delta_0 \to \Delta_1$. Similarly working out the other possibilities yields

$$\dot{t}_0 = \pi^- t_1^- - \pi^+ t_0^+, \tag{8.1a}$$
$$\dot{t}_1 = \pi^+ t_0^+ + \pi^- t_2^- - \pi^- t_1^- - \pi^+ t_1^+, \tag{8.1b}$$
$$\dot{t}_2 = \pi^+ t_1^+ + \pi^- t_3^- - \pi^- t_2^- - \pi^+ t_2^+, \tag{8.1c}$$
$$\dot{t}_3 = \pi^+ t_2^+ - \pi^- t_3^-. \tag{8.1d}$$

Let us analyse the stationary states of system (8.1a), (8.1b), (8.1c) and (8.1d). First observe that in a stationary state, the proportion of triads remains constant, so that the proportion of positive links ρ should also remain constant. Hence $\dot{\rho} = \pi^+ - \pi^- = 0$ and $\pi^+ = \pi^-$. Using $\pi^+ = \pi^-$, we obtain that $t_0^+ = t_1^-$, $t_1^+ = t_2^-$ and $t_2^+ = t_3^-$. By forming products $t_0^+ t_2^- = t_1^+ t_1^-$ and so forth, we obtain $3t_0 t_2 = t_1^2$ and $3t_1 t_3 = t_2^2$. Furthermore, because $\pi^+ = \pi^-$ we obtain that $(1-2p)t_1 = t_3$. We then obtain

$$t_3 = q^3 t_0,$$
$$t_2 = 3q^2 t_0,$$
$$t_1 = 3q t_0,$$

where $q = \sqrt{3(1-2p)}$ for $p < 1/2$ and $q = 0$ for $p \geq 1/2$. With the normalization $\sum_k t_k = 1$ we arrive at

$$t_k = \binom{3}{k} \rho_*^{3-k} (1 - \rho_*)^k, \tag{8.2}$$

where

$$\rho_* = \begin{cases} \frac{1}{q+1} & \text{for } p < 1/2 \\ 1 & \text{for } p \geq 1/2 \end{cases} \tag{8.3}$$

is the stationary proportion of friendly links ρ.

Notice that the proportion of triads are distributed according to a simple binomial distribution for $p < 1/2$. Hence, for infinitely large networks, this model converges to a distribution of triads and not to social balance for $p < 1/2$. Notice that for finite size networks it will necessarily converge to a socially balanced state, since the model runs until there are no longer any unbalanced triads. Since a balanced state is reached with non-zero probability, we are sure that after waiting long enough, the model should reach social balance. However, the analysis of large n shows that the model may spend much time in a quasi-stationary state around the binomial distribution.

For $p > 1/2$ the model will always converge to social balance, but consisting only of a single faction, and all links will be positive. Simulations also show this transition around $p \approx 1/2$.

Obtaining non trivial social balance (i.e. not consisting of a single faction) using Local Triad Dynamics (LTD) is therefore not straightforward. Although for a finite size, it will always end up in a socially balanced state in the end, this may take a very long time (about e^{n^2} from simulations [2]), and the system is expected to spend much time in a quasi-stationary state around an uncorrelated distribution of triads. Perhaps the Constrained Triad Dynamics (CTD) works better in this regard, since it only flips signs if it improves social balance. Hence, social balance can only increase in CTD. We will now investigate that model to some extent.

8.1.2 Constrained Triad Dynamics

Because in the model using Constrained Triad Dynamics (CTD) the updates depend on whether a specific link improves social balance or not, it is rather difficult to model its dynamics, also in the limit of large n. Instead, the focus here is on so-called jammed states: configurations of positive and negative links such that no sign can be flipped in order to improve social balance. Hence, the CTD becomes stuck in such a local maximum, as it has no way of changing any link. Nonetheless, it was found that, even though these jammed states exists, the dynamics often still converge to social balance, without being stuck in a jammed stated. Moreover, it does so more quickly then LTD.

We will discern two types of jammed states, following [9], strict, and weak jammed states. First let us define the energy U of a configuration.

Definition 8.1 Let A be a complete signed adjacency network (without self loops). We then define the energy as

$$U(A) = \frac{1}{\binom{n}{3}} \sum_{ijk} A_{ij} A_{jk} A_{ki} = \frac{1}{\binom{n}{3}} \operatorname{Tr} A^3.$$

This notion of energy simply counts how many triads are balanced (so have a positive sign) compared to the total number of triads $\binom{n}{3}$. Clearly in balanced networks all triads have positive sign, and so $U = 1$. A network that consists only of negative links clearly has $U = -1$. We can now define strict and weak jammed states.

Definition 8.2 Let A be a complete signed adjacency network. Let A'_{ij} be the signed adjacency matrix with sign $A'_{ij} = -A_{ij}$ flipped for link ij. We call A a strict jammed stated whenever A is unbalanced and for all i, j, we have $U(A) > U(A')$ and a weak jammed state if $U(A) \geq U(A')$.

Notice that this is equivalent to saying that each edge ij must have

$$A_{ij} \sum_k A_{jk} A_{ki} \geq -A_{ij} \sum_k A_{jk} A_{ki}. \tag{8.4}$$

since otherwise flipping ij would improve the energy U. This is equivalent to saying that more than half of the triads in which the ij participates should be balanced. It can then be easily seen that no jammed state can have $U < 0$.

Lemma 8.3 *Let A be a jammed state. Then $U(A) \geq 0$.*

Proof For all edges ij we have that

$$A_{ij} \sum_k A_{jk} A_{ki} \geq 0,$$

otherwise $A_{ij} \sum_k A_{jk} A_{ki} \geq -A_{ij} \sum_k A_{jk} A_{ki}$ and flipping ij would improve U. Hence

$$U(A) = \frac{1}{\binom{n}{3}} \sum_{ijk} A_{ij} A_{jk} A_{kj} \tag{8.5}$$

$$= \frac{1}{\binom{n}{3}} \sum_{ij} A_{ij} \sum_k A_{jk} A_{kj} \geq 0. \tag{8.6}$$

□

Notice that hence also $\mathrm{Tr}(A) \geq 0$ for jammed states. Hence, when looking for jammed states, we know that if $U(A) < 0$ the state is definitely not jammed.

Strict jammed states only exist for $n = 9$ and $n \geq 11$ and jammed states only exist for $n = 6$ or $n \geq 8$, we will only show the former. For that we first need the following observation.

Lemma 8.4 *Let $T_1 = (i, j, k)$ be an unbalanced triad and $T_2 = (i, v, k)$ be a balanced triad. Then either $T_3 = (i, j, v)$ or $T_4 = (j, v, k)$ is unbalanced.*

Proof

$$\begin{aligned} \mathrm{sgn} T_3 \mathrm{sgn} T_4 &= A_{ij} A_{jv} A_{vi} A_{jv} A_{vk} A_{kj} \\ &= \mathrm{sgn} T_1 \mathrm{sgn} T_2 = -1. \end{aligned}$$ □

Theorem 8.5 *Strict jammed states only exist for $n = 9$ and $n \geq 11$.*

Proof Let A be a jammed state of $n = 2r$ an even number of nodes with inbalanced triad (i, j, k). Let (i, k, v_s) be a balanced triad, of which at least $q \geq r$ of the $2r - 2$ triads are balanced [otherwise we could flip the sign of (i, k)]. By the previous lemma, for each v_s either (i, j, v_s) or (j, v_s, k) is unbalanced. Denote by x the number of times (i, j, v_s) is unbalanced, and by y the number of times y is unbalanced. Then $x + y = q$. Since there can be at most r triads unbalanced for the edge (i, j) (otherwise we could flip its sign), we also have that $x + 1 \leq r - 2$. Similarly, $y + 1 \leq r - 2$, and hence $x + y + 2 \leq 2r - 4$ or $r \geq 6$ so that $n \leq 12$. For $n = 2r + 1$ an odd number of nodes we obtain similarly $n \leq 9$. □

Furthermore, it can be proven that if the positive graph G^+ is a Payley graph, and the remaining edges are negative, it is a jammed state [9].

We can hence conclude that the CTD does not necessarily converge towards a socially balanced state, because of these jammed states. Nonetheless, it was observed by simulations that the probability to get stuck in a jammed state approaches 0 for large n.

8.2 Continuous Time Squared Model

As shown in the previous section, the discrete model shows some difficulties in attaining social balance. Using local triad dynamics, the system spends a large amount of time in a quasi stationary state in which the triad densities are uncorrelated. Using constrained triad dynamics, the system might be stuck in jammed states. It was suggested that by using continuous time dynamics, one would obtain a model that would always converge to social balance [8].

The suggested model has the following form. We denote by $X_{ij} \in \mathbb{R}$ the relationship between i and j, such that when $X_{ij} < 0$ the two nodes i and j are enemies, and if $X_{ij} > 0$ the two are friends. For simplicity, we assume a complete graph, including self-loops, so that the matrix X is complete. The relationships are assumed to change according to

$$\dot{X}_{ij} = \sum_k X_{ik} X_{kj} \quad \text{or } \dot{X} = X^2, \tag{8.7}$$

where $\dot{X}$ represents the time derivative. The idea behind this model is that reputations are adjusted based on the outcome of a particular gossiping process. More specifically, suppose that Bob (individual i) wants to revise his opinion about John (individual j). Bob then asks everybody else in the network what they think of John. If one such opinion X_{kj} has the same sign as the opinion Bob has about his gossiping partner, i.e. as X_{ik}, then Bob will be increase his opinion about John. But if these opinions differ in sign, then Bob will decrease his opinion about John.

The fundamental question is whether or not the solutions of Eq. (8.7) evolve towards a state which corresponds to a balanced network. As usual, we are only interested in the signs of the links, not the weights themselves. This model has the tendency to blow-up, as we will see, and so we normalize in order to facilitate the analysis, and we study

$$\lim_{t \to t^*} \frac{X(t)}{|X(t)|_F}, \tag{8.8}$$

where $|X|_F = \sqrt{\operatorname{Tr} XX^\top}$ denotes the Frobenius norm. The Frobenius norm is unitarily invariant so that $|UXU^\top|_F = |X|_F$ for a unitary matrix U (i.e. $UU^\top = I$ with I the identity). If this limit is socially balanced, the system attains a socially balanced state. Recall from Theorem 7.4 that X is balanced if and only $X = uv^\top$ for some u and v such that $u_i v_i > 0$. Hence, $X(t)$ will be balanced if and only if

$$\lim_{t \to t^*} \frac{X(t)}{|X(t)|_F} = uv^\top \tag{8.9}$$

for some u and v with $u_i v_i > 0$. In particular, if $u = v$ then $X(t)$ is balanced. Notice that if $u \neq v$ then only the weights can be different, but that the signs must be the same. Hence, we could also say that $X(t)$ is balanced if and only if

$$\lim_{t \to t^*} \operatorname{sgn} \frac{X(t)}{|X(t)|_F} = uu^\top. \tag{8.10}$$

It will be convenient to consider the decomposition of X in a symmetric and skew-symmetric part, $X = S + A$ with $S = S^\top$ and $A = -A^\top$. The symmetric and anti-symmetric part can be uniquely defined as

$$S = \frac{X + X^\top}{2}, \quad A = \frac{X - X^\top}{2}. \tag{8.11}$$

Notice that $\langle S, A\rangle = \operatorname{Tr} SA^\top = -\operatorname{Tr} SA^\top = 0$, so that $A \perp S$. We denote by $\mathcal{S}$ and $\mathcal{A}$ the set of symmetric and skew-symmetric matrices respectively. Furthermore, we denote by I_n the $n \times n$ identity matrix, and by J_n a specific skew symmetric matrix:

$$J_n = \begin{pmatrix} 0 & I_{n/2} \\ -I_{n/2} & 0 \end{pmatrix}, \quad n \text{ even.} \tag{8.12}$$

Now let us look under what conditions social balance is attained.

8.2.1 Normal Initial Condition

We start by defining

$$\mathcal{N} = \{X \in \mathbb{R}^{n\times n} | XX^\top = X^\top X\},$$

the set of real, normal matrices. Notice that $\frac{\partial XX^\top}{\partial t} = \frac{\partial X^\top X}{\partial t}$ if $X(0) \in \mathcal{N}$ so that the set $\mathcal{N}$ is invariant for $\dot{X} = X^2$. Furthermore, a symmetric matrix $X = X^\top$ is obviously normal, which extends the analysis of Marvel et al. [10] to that of normal matrices.

Recall that normal matrices are (block)-diagonalizable with blocks of size at most 2 by an orthogonal transformation: if $X_0 \in \mathcal{N}$, then

$$U^\top X_0 U = \Lambda_0, \tag{8.13}$$

where Λ_0 consists of real 1×1 scalar blocks A_i and real 2×2 blocks $B_j = \alpha_j I_2 + \beta_j J_2$ with $\beta_j \neq 0$.

Note that if $\Lambda(t)$ is the solution to the initial value problem $\dot{\Lambda} = \Lambda^2$, $\Lambda(0) = \Lambda_0$, then $X(t) := U\Lambda(t)U^\top$ is the solution to Eq. (8.7). This shows it is sufficient to solve system Eq. (8.7) in case of scalar X or in case of a specific, 2×2, normal matrix X. The scalar case is easy to solve: the solution of $\dot{x} = x^2$, $x(0) = x_0$, is

$$x(t) = \frac{x_0}{1 - x_0 t}, \tag{8.14}$$

which is easily verified. For $x_0 < 0$ this is valid for time $t \in [0, \infty)$ while for $x_0 > 0$ this is valid for time $t \in [0, 1/x_0)$ and it blows up at time $t^* = 1/x_0$, while for $x_0 = 0$, $x(t) = 0$. We turn to the 2×2 case by considering:

$$\dot{X} = X^2, \;\; X(0) = \alpha I_2 + \beta J_2, \text{ where } \beta > 0. \tag{8.15}$$

Lemma 8.6 *The forward solution* $X(t)$ *of* Eq. (8.15) *is defined for all* $t \in [0, +\infty)$, *and*

$$\lim_{t\to+\infty} X(t) = 0 \textit{ and } \lim_{t\to+\infty} \frac{X(t)}{|X(t)|_F} = -\frac{\sqrt{2}}{2} I_2.$$

Proof Let $X_0 = S_0 + A_0$, $S_0 = \alpha I_2$ and $A_0 = \beta J_2$ where J_2 is as defined in Eq. (8.12). Then the solution $X(t)$ of Eq. (8.15) can be decomposed as $S(t) + A(t)$, where

$$\dot{S} = S^2 + A^2, \quad S(0) = S_0, \tag{8.16a}$$

$$\dot{A} = AS + SA, \quad A(0) = A_0. \tag{8.16b}$$

Note that Eq. (8.16a) is a matrix Riccati differential equation [1] with the property that the set $\mathcal{L} := \{sI_2 + aJ_2 | s,\ a \in \mathbb{R}\}$, is an invariant set under the flow. Therefore it suffices to solve the scalar Riccati differential equation corresponding to the dynamics of the scalar coefficients s and a:

$$\dot{s} = s^2 - a^2, \quad s(0) = \alpha, \tag{8.17a}$$

$$\dot{a} = 2as, \quad a(0) = \beta, \tag{8.17b}$$

whose solution is given implicitly by:

$$s^2 + \left(a - \frac{1}{2c}\right)^2 = \left(\frac{1}{2c}\right)^2 \quad \text{if } c \neq 0,$$

where c is an integration constant. So, the orbits form circles which are centred at $(0, 1/2c)$ and pass through $(0, 0)$, and by $a = 0$ if $c = 0$. The phase portrait of system (8.17) is illustrated in Fig. 8.1.

All solutions $(s(t), a(t))$ of system (8.17), not starting on the s-axis, converge to zero as $t \to +\infty$, and approach the origin in the second quadrant for solutions in the upper-half-plane, and in the third quadrant for solutions in the lower-half-plane. Moreover, since the s-axis is the tangent line to every circular orbit at the origin, the slopes $a(t)/s(t)$ converge to 0 along every solution $\lim_{t\to+\infty} a(t)/s(t) = 0$. Consequently, the forward solution $X(t)$ of Eq. (8.15) satisfies:

$$\lim_{t\to+\infty} X(t) = \lim_{t\to+\infty} s(t)I_2 + a(t)J_2 = 0,$$

and

$$\lim_{t\to+\infty} \frac{X(t)}{|X(t)|_F} = -\frac{\sqrt{2}}{2} I_2. \qquad \square$$

Combining the solution for the scalar and 2×2 case yields our main result in the normal case:

Theorem 8.7 *Let* $X_0 \in \mathcal{N}$, *and let* (U, Λ_0) *be as in* Eq. (8.13). *Define*

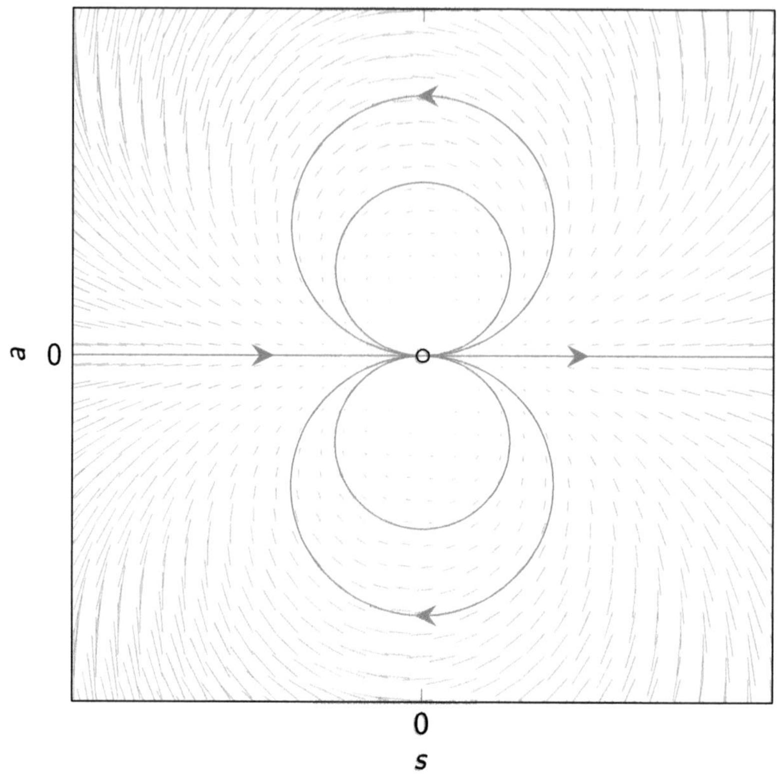

Fig. 8.1 Phase portrait of complex eigenvalue

$$\bar{t}_i = \begin{cases} 1/a_i & \textit{if } a_i > 0 \\ +\infty & \textit{if } a_i \leq 0 \end{cases} \quad \textit{for all } i = 1, \ldots, k,$$

where a_i correspond to a real eigenvalue, and let $\bar{t} = \min_i \bar{t}_i$. Then the forward solution $X(t)$ of Eq. (8.7) *is defined for* $[0, \bar{t})$.

If there is a unique $i^* \in \{1, \ldots, k\}$ such that $\bar{t} = \bar{t}_{i^*}$ is finite, then

$$\lim_{t \to \bar{t}_{i^*}-} \frac{X(t)}{|X(t)|_F} = U_{i^*} U_{i^*}^\top,$$

where U_{i^*} is the i^*th column of U, an eigenvector corresponding to eigenvalue a_{i^*} of X_0.

Proof Consider the initial value problem:

$$\dot{\Lambda} = \Lambda^2, \quad \Lambda(0) = \Lambda_0.$$

Its solution is given by

$$\Lambda(t) = \begin{pmatrix} \frac{a_1}{1-a_1 t} & \dots & 0 & 0 & \dots & 0 \\ \vdots & \ddots & \vdots & \vdots & \ddots & \vdots \\ 0 & \dots & \frac{a_k}{1-a_k t} & 0 & \dots & 0 \\ 0 & \dots & 0 & X_1(t) & \dots & 0 \\ \vdots & \ddots & \vdots & \vdots & \ddots & \vdots \\ 0 & \dots & 0 & 0 & \dots & X_l(t) \end{pmatrix},$$

where for all $j = 1, \dots, l$, $X_j(t)$ is the forward solution of Eq. (8.15), which is defined for all t in $[0, +\infty)$, and converges to 0 as $t \to +\infty$ by Lemma 8.6.

This clearly shows that $\Lambda(t)$ is defined in forward time for t in $[0, \bar{t})$. Since the solution of Eq. (8.7) is given by $X(t) = U\Lambda(t)U^\top$, $X(t)$ is also defined in forward time for t in $[0, \bar{t})$. It follows from unitary invariance of the Frobenius norm that

$$\frac{X(t)}{|X(t)|_F} = U\frac{\Lambda(t)}{|\Lambda(t)|_F}U^\top.$$

If $i^* \in \{1, \dots, k\}$ is the unique value such that $\bar{t} = \bar{t}_{i^*}$, then using the unitary invariance

$$\lim_{t\to\bar{t}_i^*} \frac{X(t)}{|X(t)|_F} = U \lim_{t\to\bar{t}_i^*} \frac{\Lambda(t)}{|\Lambda(t)|_F}U^\top = Ue_{i^*}e_{i^*}^\top U^\top = U_{i^*}U_{i^*}^\top,$$

where e_{i^*} denotes the i^*th standard unit basis vector of $\mathbb{R}^n$. □

Theorem 8.7 provides a sufficient condition guaranteeing that social balance is achieved as stated in Theorem 7.4.

8.2.2 Generic Initial Condition

Although Theorem 8.7 provides a sufficient condition for the emergence of social balance, it requires that the initial condition X_0 is normal. But the set $\mathcal{N}$ of normal matrices has measure zero in the set of all real $n \times n$ matrices, and thus the question arises if social balance will arise for non-normal initial conditions as well. We investigate this issue here, and will see that generically, social balance is not achieved.

If X_0 is a general real $n \times n$ matrix, we can put it in real Jordan canonical form by means of a similarity transformation:

$$X(0) = T\Lambda_0 T^{-1}, \quad TT^{-1} = I_n, \tag{8.18}$$

with $\Lambda_0 = \text{diag}(A_1, \ldots, A_k, B_1, \ldots, B_l)$, where A_i are real Jordan blocks and

$$B_j = \begin{pmatrix} C_i & I_2 & \ldots & 0 \\ 0 & C_i & \ddots & 0 \\ \vdots & \vdots & \ddots & \vdots \\ 0 & 0 & \ldots & C_i \end{pmatrix}, \quad C_j = \alpha_j I_2 + \beta_j J_2, \tag{8.19}$$

with $\beta_j \neq 0$. The analysis could be done similarly using complex eigenvalues, but we prefer to remain in the real domain.

We again observe that if $\Lambda(t)$ is the solution to the initial value problem $\dot{\Lambda} = \Lambda^2$, $\Lambda(0) = \Lambda_0$, then $X(t) := T\Lambda(t)T^{-1}$, is the solution to Eq. (8.7). Again, it is sufficient to solve system (8.7) in case of specific block-triangular X of the form A_i or B_j as in Eq. (8.19). To deal with the first form A_i, we first we consider more general, triangular Toeplitz initial conditions:

$$X(0) = \begin{pmatrix} x_1(0) & x_2(0) & \cdots & x_n(0) \\ 0 & x_1(0) & \ddots & x_{n-1}(0) \\ \vdots & \vdots & \ddots & \vdots \\ 0 & 0 & \cdots & x_1(0) \end{pmatrix}, \tag{8.20}$$

with $x_i(0)$ reals, and denote $\mathcal{TT} = \{X | X \text{is of the form } (8.20)\}$. It turns out that this is an invariant set for the system, which can be easily verified by noting that if X belongs to $\mathcal{TT}$, then so does X^2.

Lemma 8.8 *Let $X(0) \in \mathcal{TT}$ with*

$$x_i(0) = \begin{cases} a \neq 0 & \textit{if } i = 1 \\ 1 & \textit{if } i = 2 \\ 0 & \textit{otherwise.} \end{cases}$$

Then the forward solution $X(t)$ of Eq. (8.7) *is defined on $[0, t^*)$ where $t^* = 1/a$ if $a > 0$ and on $t^* = \infty$ if $a \leq 0$, belongs to $\mathcal{TT}$, and satisfies*

$$x_i(t) = p_i\left(\frac{1}{1-at}\right), \quad t \in [0, t^*),$$

where each $p_i(z)$ is a polynomial of degree i:

$$p_i(z) = \begin{cases} az & \textit{if } i = 1 \\ \frac{1}{a^{i-2}} z^i + \cdots + c_i z^2 & \textit{otherwise,} \end{cases} \tag{8.21}$$

where c_i is some real constant, so that $p_i(z)$ has no constant or first order terms when $i > 1$.

Proof First note that system (8.7) can be solved recursively for matrices of type (8.20), starting with $x_1(t)$, followed by $x_2(t), x_3(t), \ldots$. Only the first equation for x_1 is non-linear, whereas the equations for $x_2, x_3, \ldots$ are linear. To see this, we write these equations:

$$\dot{x}_i = \begin{cases} x_1^2,\ x_1(0) = a & \text{if } i = 1 \\ (2x_1(t))x_2,\ x_2(0) = 1 & \text{if } i = 2 \\ (2x_1(t))x_i + \sum_{k=2}^{i-1} x_k(t)x_{i-(k-1)}(t),\ x_i(0) = 0 & \text{if } i > 2. \end{cases}$$

The forward solution for x_1 is $x_1(t) = \frac{a}{1-at}$, for $t \in [0, t^*)$, which establishes the result if $i = 1$. The forward solution for x_2 is: $x_2(t) = \frac{1}{(1-at)^2}$, for $t \in [0, t^*)$, which establishes the result if $i = 2$. If $i > 2$, we obtain the proof by induction on n. Assume the result holds for $i = 1, \ldots, n$, for some $n \geq 2$, and consider the equation for x_{n+1}. Using that $x_n(0) = 0$ for $n \geq 2$, the solution is given by:

$$x_{n+1}(t) = e^{\int_0^t 2x_1(s)ds}\left[0 + \int_0^t \left(\sum_{k=2}^{n} x_k(s)x_{n-k+2}(s)\right) e^{\int_0^s -2x_1(\tau)d\tau} ds\right].$$

Since $e^{\int_0^t 2x_1(s)ds} = x_2(t)$ and thus $e^{\int_0^s -2x_1(\tau)d\tau} = 1/x_2(s)$, it follows that:

$$x_{n+1}(t) = \frac{1}{(1-at)^2}\left[\int_0^t \left(\sum_{k=2}^{n} p_k(1/(1-as)) \right.\right.$$
$$\left.\left. p_{n-k+2}(1/(1-as))\right)(1-as)^2 ds\right].$$

Since the polynomials appearing in the integral take the form of Eq. (8.21), they are all missing first order and constant terms, and thus there follows that

$$x_{n+1}(t) = \frac{1}{(1-at)^2}\left[\int_0^t \left(\sum_{k=2}^{n} \frac{1}{a^{n-2}}\frac{1}{(1-as)^{n+2}} \right.\right.$$
$$\left.\left. + \cdots + c_k c_{n-k+2}\frac{1}{(1-as)^4}\right)(1-as)^2 ds\right]$$

and so that

$$x_{n+1}(t) = \frac{1}{a^{n-1}}\frac{1}{(1-at)^{n+1}} + \cdots + \frac{c_{n+1}}{(1-at)^2}, \quad t \in [0, t^*),$$

where K_{n+1} and c_{n+1} are certain constants (which are related in some way which is irrelevant for what follows). This shows that $x_{n+1}(t)$ is indeed of the form $p_{n+1}(1/(1-at))$ with $p_{n+1}(z)$ as in Eq. (8.21). □

Next we consider Eq. (8.7) in case $X(0)$ is a block triangular Toeplitz initial condition:

$$X(0) = \begin{pmatrix} B_1(0) & B_2(0) & \cdots & B_n(0) \\ 0 & B_1(0) & \ddots & B_{n-1}(0) \\ \vdots & \vdots & \ddots & \vdots \\ 0 & 0 & \cdots & B_1(0) \end{pmatrix}, \tag{8.22}$$

with $B_i(0) = \alpha_i I_2 + \beta_i J_2$ with $\alpha_i, \beta_i \in \mathbb{R}$, and denote

$$\mathcal{BTT} = \{X | X \text{is of the form } (8.22)\}.$$

Again the set $\mathcal{BTT}$ is invariant for system (8.7). We use this to solve Eq. (8.7) in case $X(0)$ is a real Jordan block corresponding to a pair of eigenvalues $\alpha \pm j\beta$.

Lemma 8.9 *Let $X(0) \in \mathcal{BTT}$ with*

$$B_i(0) = \begin{cases} \alpha I_2 + \beta J_2 & \text{if } i = 1 \\ I_2 & \text{if } i = 2 \\ 0 & \text{otherwise.} \end{cases}$$

Then the forward solution $X(t)$ of Eq. (8.7) *is defined on* $[0, +\infty)$, *and it belongs to $\mathcal{BTT}$.*

Proof Just like in the proof of Proposition 8.8, we note that system (8.7) can be solved recursively, starting with $X_1(t)$, followed by $X_2(t)$, $X_3(t)$, ... Only the first equation for X_1 is non-linear, whereas the equations for $X_2, X_3, \ldots$ are linear. To see this, we write these equations

$$\dot{X}_i = \begin{cases} X_1^2, \; X_1(0) = \alpha I_2 + \beta J_2 & \text{if } i = 1 \\ (2X_1(t))X_2, \; X_2(0) = I_2 & \text{if } i = 2 \\ (2X_1(t))X_i + \sum_{k=2}^{i-1} X_k(t) X_{i-(k-1)}(t), \; X_i(0) = 0 & \text{if } i > 2. \end{cases}$$

Here we have used the fact that $X_1 X_i + X_i X_1 = 2X_1 X_i$, since any two matrices of the form $pI_2 + qJ_2$ commute and the matrices $X_i(t)$ are of this form.

By Lemma 8.6, the forward solution for $X_1(t)$ is defined for all t in $[0, +\infty)$ (and in fact, converges to zero as $t \to +\infty$).

Since the $X_1(t)$ commute for every pair of t's, the forward solution for $X_2(t)$ is given by Rugh [12] $X_2(t) = e^{\int_0^t 2X_1(s)ds}$, for $t \in [0, +\infty)$, where this solution exists for all forward times t because $X_1(t)$ is bounded and continuous. Similarly, the forward solution for $X_i(t)$ when $i > 2$, is given by the variation of constants formula:

$$X_i(t) = X_2(t)\left[\int_0^t X_2^{-1}(s)\left(\sum_{k=2}^{i-1} X_k(s)X_{i-(k-1)}(s)\right)ds\right],$$

for $t \in [0, +\infty)$ when $i > 2$, where these solutions are recursively defined for all forward times because the formula only involves integrals of continuous functions. □

Combining both results puts us in a position to state and prove our main result.

Theorem 8.10 *Let* $X(0) \in \mathbb{R}^{n\times n}$ *and* (T, Λ_0) *as in* Eq. (8.18) *with* (8.19). *Let* $a_1 > a_2 \geq \cdots \geq a_k$ *with* $a_1 > 0$ *a simple eigenvalue with corresponding right and left-eigenvectors* U_1 *and* $V_1^{\top}$ *respectively:*

$$X(0)U_1 = a_1U_1 \quad \text{and } V_1^{\top}X(0) = a_1V_1^{\top}.$$

Then the forward solution $X(t)$ *of* Eq. (8.7) *is defined for* $[0, 1/a_1)$, *and*

$$\lim_{t\to 1/a_1} \frac{X(t)}{|X(t)|_F} = \frac{U_1V_1^{\top}}{|U_1V_1^{\top}|_F}.$$

Proof Consider the initial value problem $\dot{\Lambda} = \Lambda^2$, $\Lambda(0) = \Lambda_0$, whose solution is given by

$$\Lambda(t) = \text{diag}(A_1(t), \ldots, A_k(t), B_1(t), \ldots, B_l(t)),$$

where for all $i = 1, \ldots, k$, $A_i(t)$ is the forward solution of Eq. (8.7) with $A_i(0)$ of the form A_i in (8.19), which by Lemma 8.8 is defined for all $t \in [0, 1/a_i)$. Since $a_1 > a_2 \geq \cdots \geq a_k$, $A_1(t)$ blows up first when $t \to 1/a_1$. The matrices $B_j(t)$, $j = 1, \ldots, l$, are the forward solution of Eq. (8.7) with $B_j(0)$ of the form B_j in Eq. (8.19), and by Lemma 8.9, they are defined for all t in $[0, +\infty)$.

This clearly shows that $\Lambda(t)$ is defined in forward time for t in $[0, 1/a_1)$. Since the solution of Eq. (8.7) is given by $X(t) = T\Lambda(t)T^{-1}$, $X(t)$ is also defined in forward time for t in $[0, 1/a_1)$, and it follows that

$$\begin{aligned}
\lim_{t\to 1/a_1} \frac{X(t)}{|X(t)|_F} &= \lim_{t\to 1/a_1} \frac{T\Lambda(t)T^{-1}}{|X(t)|_F} \\
&= \frac{Te_1e_1^{\top}T^{-1}}{|Te_1e_1^{\top}T^{-1}|_F} = \frac{U_1V_1^{\top}}{|U_1V_1^{\top}|_F},
\end{aligned}$$

where e_1 denotes the first standard unit basis vector of $\mathbb{R}^n$. □

Theorem 8.10 implies that social balance is usually not achieved when $X(0)$ is an arbitrary real initial condition, illustrated in Fig. 8.2. Indeed, if X_0 has a simple, positive, real eigenvalue a_1, and if we assume that no entry of the right and left

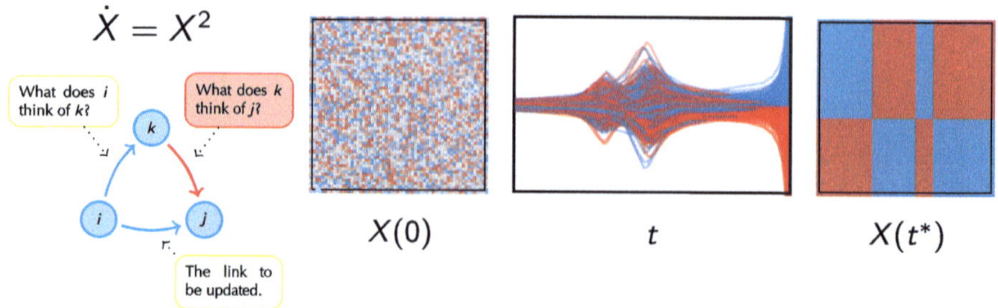

Fig. 8.2 Generic behaviour of $\dot{X} = X^2$

eigenvectors U_1 and $V_1^\top$ are zero (an assumption which is generically satisfied), then in general, up to a permutation of its entries, the sign patterns of U_1 and $V_1^\top$ are:

$$U_1 = \begin{pmatrix} + \\ + \\ \hline - \\ - \end{pmatrix} \text{ and } V_1^\top = (+\ -\,|\,+\ -)$$

implies that

$$U_1 V_1^\top = \left(\begin{array}{cc|cc} + & - & + & - \\ + & - & + & - \\ \hline - & + & - & + \\ - & + & - & + \end{array}\right).$$

Then Theorem 7.4 implies that the normalized state of the system does not become balanced in finite time.

This shows that in general, unless X_0 is normal (so that Theorem 8.7 is applicable), we cannot expect that social balance will emerge for the model $\dot{X} = X^2$.

8.3 Continuous Time Transpose Model

The previous model $\dot{X} = X^2$ will in general not converge to social balance, unless the initial condition is normal with a positive single real eigenvalue. Hence, for general initial conditions, we do not expect the model to converge to social balance. Is remains open to see what model then is expected to converge to social balance in general. In this section we will suggest such a model, and prove that generically it converges to social balance.

Let us briefly reconsider the gossiping process underlying the model $\dot{X} = X^2$. In our example of Bob and John, the following happens. Bob asks others what they think of John. Bob takes into account what he thinks of the people he talks to, and adjusts his opinion of John accordingly. An alternative approach is to consider a type of homophily process [5, 6, 11]: people tend to befriend people who think alike. When Bob seeks to revise his opinion of John, he talks to John about everybody else (instead of talking to everybody else about John). For example, suppose that Bob likes Alice, but that John dislikes her. When Bob and John talk about Alice, they notice they have opposing views about her, and as a result the relationship between Bob and John deteriorates. On the other hand, should they share similar opinions about Alice, their relationship will improve. Thus, our alternative model for the update law of the reputations is:

$$\dot{X}_{ij} = \sum_k X_{ik} X_{jk} \quad \text{or } \dot{X} = XX^\top, \ X(0) = X_0, \tag{8.23}$$

where again, each X_{ij} denotes the real-valued opinion agent i has about agent j. Notice that for $i = j$, the value of X_{ii} is interpreted as a measure of self-esteem of agent i. In this case $\dot{X}_{ii} = \sum_k X_{ik}^2 \geq 0$ contrary to the earlier model, and so the self-esteem is always increasing. We shall call this model the "transpose model", because of the transpose we use in the model. Notice that an alternative specification could be $\dot{X} = X^\top X$, in which case Bob and John don't talk about what they think of Alice, but what Alice thinks of them. Although this seems to be less realistic, it might be an interesting model to study nonetheless. In this thesis we will limit ourselves to the model $\dot{X} = XX^\top$.

As in the case of model $\dot{X} = X^2$, we split up the analysis in two parts. First we consider system (8.23) with normal initial condition X_0, and we shall see that not all initial conditions lead to the emergence of a balanced network in this case, in contrast to the behaviour of Eq. (8.7). Secondly, we will see that for non-normal, generic initial conditions X_0, we typically do get the emergence of social balance, also contrasting the behaviour of Eq. (8.7).

8.3.1 Normal Initial Condition

As for the model $\dot{X} = X^2$ the set $\mathcal{N}$ of normal matrices is invariant for $\dot{X} = XX^\top$. By using the same diagonalisation as in Eq. (8.13), if $\Lambda(t)$ is the solution to the initial value problem $\dot{\Lambda} = \Lambda\Lambda^\top$, $\Lambda(0) = \Lambda_0$, then $X(t) := U\Lambda(t)U^\top$, is the solution to Eq. (8.23). This shows it is sufficient to solve system (8.23) in case of scalar X or in case of a specific 2×2 normal matrix X. The scalar case is easy to solve and follows Eq. (8.14), so we turn to the 2×2 case by considering

$$\dot{X} = XX^\top, \ X(0) = \alpha I_2 + \beta J_2, \quad \text{where } \beta \neq 0. \tag{8.24}$$

We define the angle ϕ as

$$\phi = \arctan\left(\frac{\alpha}{\beta}\right), \ \phi \in \left(-\frac{\pi}{2}, \frac{\pi}{2}\right). \tag{8.25}$$

Lemma 8.11 *Define* $\bar{t}$ *as*

$$\bar{t} = \frac{\pi}{2\beta} - \frac{\phi}{\beta}. \tag{8.26}$$

Then the forward solution $X(t)$ *of* Eq. (8.24) *is:*

$$X(t) = \beta \tan(\beta t + \phi) I_2 + \beta J_2, \quad t \in [0, \bar{t}). \tag{8.27}$$

Moreover,

$$\lim_{t \to \bar{t}-} X(t) = +\infty I_2 + \beta J_2 \text{ and } \lim_{t \to \bar{t}-} \frac{X(t)}{|X(t)|_F} = \frac{\sqrt{2}}{2} I_2.$$

Proof Let $X_0 = S_0 + A_0$, $S_0 = \alpha I_2$, and $A_0 = \beta J_2$. Then the solution $X(t)$ of Eq. (8.24) can be decomposed as $S(t) + A(t)$, where

$$\dot{S} = (S + A)(S - A), \quad S(0) = S_0, \tag{8.28a}$$

$$\dot{A} = 0, \quad A(0) = A_0, \tag{8.28b}$$

so $A(t) = A_0$, and reduces to

$$\dot{S} = (S + A_0)(S - A_0), \quad S(0) = S_0 \tag{8.29}$$

Note that Eq. (8.29) is a matrix Riccati differential equation with the property that the line $\mathcal{L} = \{\alpha I_2 | \alpha \in \mathbb{R}\}$, is an invariant set under the flow. Therefore it suffices to solve the scalar Riccati differential equation corresponding to the dynamics of the diagonal entries of S: $\dot{s} = s^2 + \beta^2$, $s(0) = \alpha$, whose forward solution is: $s(t) = \beta \tan(\beta t + \phi)$, for $t \in (0, \bar{t})$, where $\bar{t}$ is given by Eq. (8.26). Consequently, the forward solution $X(t)$ of Eq. (8.24) is given by: $X(t) = S(t) + A_0 = \beta \tan(\beta t + \phi) I_2 + \beta J_2$, for $t \in (0, \bar{t})$, and thus $\lim_{t \to \bar{t}-} X(t) = +\infty I_2 + \beta J_2$ and

$$\lim_{t \to \bar{t}-} \frac{X(t)}{|X(t)|_F} = \frac{X(t)}{\sqrt{2} |\beta \sec(\beta t + \phi)|} = \frac{\sqrt{2}}{2} I_2. \qquad \square$$

Combining the solution for the 1×1 scalar case in Eq. (8.14) and Lemma 8.11 yields our main result:

Theorem 8.12 *Let* $X_0 \in \mathcal{N}$, *and let* (U, Λ_0) *be as in* Lemma 8.13. *Define*

$$\bar{t}_i = \begin{cases} 1/a_i & \text{if } a_i > 0 \\ +\infty & \text{if } a_i \leq 0 \end{cases} \quad \text{for all } i = 1, \ldots, k,$$

and

$$\bar{t}_j = \frac{\pi}{2\beta_j} - \frac{\phi_j}{\beta_j} \quad \textit{for all} \;\; j = 1, \ldots, l,$$

where $\phi_j = \arctan\left(\frac{\alpha_j}{\beta_j}\right)$ *and let* $\bar{t} = \min_{i,j}\{\bar{t}_i, \bar{t}_j\}$. *Then the forward solution* $X(t)$ *of* Eq. (8.23) *is defined for* $[0, \bar{t})$.

If there is a unique $i^* \in \{1, \ldots, k\}$ *such that* $\bar{t} = \bar{t}_{i^*}$ *is finite, then*

$$\lim_{t \to \bar{t}_{i^*}-} \frac{X(t)}{|X(t)|_F} = U_{i^*} U_{i^*}^\top,$$

where U_{i^*} *is the* i^**th column of* U, *an eigenvector corresponding to eigenvalue* a_{i^*} *of* X_0.

If there is a unique $j^* \in \{1, \ldots, l\}$ *such that* $\bar{t} = \bar{t}_{j^*}$, *then*

$$\lim_{t \to \bar{t}_{j^*}-} \frac{X(t)}{|X(t)|_F} = \frac{\sqrt{2}}{2} U_{j^*} U_{j^*}^\top,$$

where U_{j^*} *is an* $n \times 2$ *matrix consisting of the two consecutive columns of* U *which correspond to the columns of the* 2×2 *block* B_{j^*} *in* Λ_0.

Proof Consider the initial value problem:

$$\dot{\Lambda} = \Lambda \Lambda^\top, \quad \Lambda(0) = \Lambda_0.$$

By Lemma 8.11 its solution is given by

$$\Lambda(t) = \begin{pmatrix} \frac{a_1}{1-a_1 t} & \cdots & 0 & 0 & \cdots & 0 \\ \vdots & \ddots & \vdots & \vdots & \ddots & \vdots \\ 0 & \cdots & \frac{a_k}{1-a_k t} & 0 & \cdots & 0 \\ 0 & \cdots & 0 & X_1(t) & \cdots & 0 \\ \vdots & \ddots & \vdots & \vdots & \ddots & \vdots \\ 0 & \cdots & 0 & 0 & \cdots & X_l(t) \end{pmatrix},$$

where for all $j = 1, \ldots, l$, $X_j(t)$ is given by the 2×2 matrix in Eq. (8.27) with β, ϕ and $\bar{t}$ replaced by β_j, ϕ_j and $\bar{t}_j$ respectively. This clearly shows that $\Lambda(t)$ is defined in forward time for t in $[0, \bar{t})$. Since the solution of Eq. (8.23) is given by $X(t) = U\Lambda(t)U^\top$, $X(t)$ is also defined in forward time for t in $[0, \bar{t})$. It follows from unitary invariance of the Frobenius norm that

$$\frac{X(t)}{|X(t)|_F} = U \frac{\Lambda(t)}{|\Lambda(t)|_F} U^\top.$$

If $i^* \in \{1, \dots, k\}$ is the unique value such that $\bar{t} = \bar{t}_{i^*}$, then

$$\begin{aligned}\lim_{t \to \bar{t}_i^*} \frac{X(t)}{|X(t)|_F} &= U \lim_{t \to \bar{t}_i^*} \frac{\Lambda(t)}{|\Lambda(t)|_F} U^\top \\ &= U e_{i^*} e_{i^*}^\top U^\top = U_{i^*} U_{i^*}^\top,\end{aligned}$$

where e_{i^*} denotes the i^*th standard unit basis vector of $\mathbb{R}^n$.

If $j^* \in \{1, \dots, l\}$ is the unique value such that $\bar{t} = \bar{t}_{j^*}$, then by Lemma 8.11:

$$\begin{aligned}\lim_{t \to \bar{t}_j^*} \frac{X(t)}{|X(t)|_F} &= U \lim_{t \to \bar{t}_j^*} \frac{\Lambda(t)}{|\Lambda(t)|_F} U^\top \\ &= \frac{\sqrt{2}}{2} U E_{j^*} U^\top = \frac{\sqrt{2}}{2} U_{j^*} U_{j^*}^\top,\end{aligned}$$

where E_{j^*} has exactly two non-zero entries equal to 1 on the diagonal positions corresponding to the block B_{j^*} in Λ_0. □

A particular consequence of Theorem 8.12 is that if X_0 has a complex pair of eigenvalues, the solution of $\dot{X} = XX^\top$ always blows up in finite time, even if all real eigenvalues of X_0 are non-positive. Recall that the solution of $\dot{X} = X^2$ blows up in finite time, if and only if X_0 has a positive, real eigenvalue. Another implication of Theorem 8.12 is that if blow-up occurs, it may be due to a real eigenvalue of X_0, or to a complex eigenvalue. In contrast, if the solution of $\dot{X} = X^2$ blows up in finite time, it is necessarily due to a positive, real eigenvalue, and never to a complex eigenvalue. When the solution of $\dot{X} = XX^\top$ blows up because of a positive, real eigenvalue of X_0, the system will achieve balance, just as in the case of system $\dot{X} = X^2$. If on the other hand, finite time blow up of $\dot{X} = XX^\top$ is caused by a complex eigenvalue of X_0, we show that in general one cannot expect to achieve a balanced network. Assume there is a unique j^* such that:

$$\lim_{t \to \bar{t}_j^*-} \frac{X(t)}{|X(t)|_F} = \frac{\sqrt{2}}{2} U_{j^*} U_{j^*}^\top,$$

Assuming that no entry of U_{j^*} is zero, the sign pattern of $U_{j^*} U_{j^*}^\top$, with

$$U_j^* = \begin{pmatrix} p_1 & q_1 \\ p_2 & -q_2 \\ -p_3 & q_3 \\ -p_4 & -q_4 \end{pmatrix}$$

is given by:

$$\begin{pmatrix} + & ? & ? & - \\ ? & + & - & ? \\ ? & - & + & ? \\ - & ? & ? & + \end{pmatrix},$$

up to a suitable permutation, where all p_i and $q_i, i = 1, \dots, 4$, are entry-wise positive vectors, and where

$$\langle p_1, q_1 \rangle + \langle p_4, q_4 \rangle = \langle p_2, q_2 \rangle + \langle p_3, q_3 \rangle,$$

because U is an orthogonal matrix. The ? are not entirely arbitrary because $U_{j*}U_{j*}^{\top}$ is a symmetric matrix, but besides that their signs can be arbitrary.

8.3.2 Generic Initial Condition

Consider

$$\dot{X} = XX^{\top}, \quad X(0) = X_0, \tag{8.30}$$

where X is a real $n \times n$ matrix, which is not necessarily normal.

We first decompose the flow (8.30) into flows for the symmetric and skew-symmetric parts of X. Let $X = S + A$, $X_0 = S_0 + A_0$, where $S, S_0 \in \mathcal{S}$ and $A, A_0 \in \mathcal{A}$ are the unique symmetric and skew-symmetric parts of X and X_0 respectively. If $X(t)$ satisfies Eq. (8.30), then it can be verified that $S(t)$ and $A(t)$ satisfy the system:

$$\dot{S} = (S + A)(S - A), \quad S(0) = 0, \tag{8.31}$$

$$\dot{A} = 0, \quad A(0) = A_0, \tag{8.32}$$

Consequently, $A(t) = A_0$ for all t, and thus the skew-symmetric part of the solution $X(t)$ of Eq. (8.30) remains constant and equal to A_0. Throughout this section we assume that $A_0 \neq 0$, for otherwise $X(0)$ is symmetric, hence normal, and the results from the previous section apply. It follows that we only need to understand the dynamics of the symmetric part. Then the solution $X(t)$ to Eq. (8.30) is given by $X(t) = S(t) + A_0$, where $S(t)$ solves Eq. (8.31), and in view of $S \perp A$, there follows by Pythagoras' Theorem that: $|X(t)|_F^2 = |S(t)|_F^2 + |A_0|_F^2$, and thus

$$\frac{X(t)}{|X(t)|_F} = \frac{S(t) + A_0}{\left(|S(t)|_F^2 + |A_0|_F^2\right)^{\frac{1}{2}}}. \tag{8.33}$$

Next we shall derive an explicit expression for the solution of Eq. (8.31). We start by performing a change of variables:

$$\hat{S}(t) = \mathrm{e}^{-tA_0}\, S(t)\, \mathrm{e}^{tA_0}\,. \tag{8.34}$$

This yields the equation

$$\dot{\hat{S}} = \hat{S}^2 - A_0^2, \quad \hat{S}(0) = S_0. \tag{8.35}$$

We perform a further transformation which diagonalizes $-A_0^2$. Let V be an orthogonal matrix such that $-V^\top A_0^2 V = D^2$, where D is the diagonal matrix $D := \mathrm{diag}(0, \omega_1 I_2, \ldots, \omega_k I_k)$ where $k \geq 1$ (because $A_0 \neq 0$) and all $\omega_j > 0$ without loss of generality. Setting

$$\tilde{S} = V^\top \hat{S} V, \tag{8.36}$$

and multiplying Eq. (8.35) by V on the left, and by $V^\top$ on the right, we find that:

$$\dot{\tilde{S}} = \tilde{S}^2 + D^2, \quad \tilde{S}(0) = \tilde{S}_0 := V^\top S_0 V. \tag{8.37}$$

Notice that this is a matrix Riccati differential equation, a class of equations with specific properties which are briefly reviewed next.

Consider a general matrix Riccati differential equation:

$$\dot{S} = SMS - SL - L^\top S + N, \tag{8.38}$$

where $M = M^\top$, $N = N^\top$ and L arbitrary, defined on $\mathcal{S}$. Associated to this equation is a linear system

$$\begin{pmatrix} \dot{P} \\ \dot{Q} \end{pmatrix} = H \begin{pmatrix} P \\ Q \end{pmatrix}, \quad H := \begin{pmatrix} L & -M \\ N & -L^\top \end{pmatrix}, \tag{8.39}$$

where H is a Hamiltonian matrix, i.e. $J_{2n} H = (J_{2n} H)^\top$ holds, where J_{2n} is as defined in Eq. (8.12). The following fact is well-known [1].

Lemma 8.13 *Let* $\begin{pmatrix} P(t) \\ Q(t) \end{pmatrix}$ *be a solution of* Eq. (8.39). *Then, provided that* $P(t)$ *is non-singular,*

$$S(t) = Q(t) P(t)^{-1}, \tag{8.40}$$

is a solution of Eq. (8.38). *Conversely, if* $S(t)$ *is a solution of* Eq. (8.38), *then there exists a solution* $\begin{pmatrix} P(t) \\ Q(t) \end{pmatrix}$ *of* Eq. (8.39) *such that* Eq. (8.40) *holds, provided that* $P(t)$ *is non-singular.*

Proof Taking derivatives in $S(t)P(t) = Q(t)$ yields that $\dot{S} = (\dot{Q} - S\dot{P})P^{-1}$, and using Eq. (8.39),

$$\dot{S} = (NP - L^\top Q - S(LP - MQ))P^{-1} = N - L^\top S - SL + SMS,$$

showing that $S(t)$ solves Eq. (8.38). For the converse, let $S(t)$ be a solution of Eq. (8.38). Let $\begin{pmatrix} P(t) \\ Q(t) \end{pmatrix}$ with $\begin{pmatrix} P(0) \\ Q(0) \end{pmatrix} = \begin{pmatrix} I_n \\ S(0) \end{pmatrix}$ be the solution of Eq. (8.39). Then

$$\begin{aligned} \frac{d}{dt}\left(Q(t)P^{-1}(t)\right) &= \dot{Q}P^{-1} - QP^{-1}\dot{P}P^{-1} \\ &= (NP - L^\top Q)P^{-1} - QP^{-1}(LP - MQ)P^{-1} \\ &= (QP^{-1})M(QP^{-1}) - (QP^{-1})L - L^\top(QP^{-1}) + N, \end{aligned}$$

implying that QP^{-1} is a solution to Eq. (8.38). Since $S(0) = Q(0)P^{-1}(0)$, it follows from uniqueness of solutions that $S(t) = Q(t)P^{-1}(t)$. □

In other words, in principle we can solve the non-linear equation (8.38) by first solving the linear system (8.39), and then use formula (8.40) to determine the solution of Eq. (8.38).

We carry this out for our particular Riccati equation (8.37) which is of the form (8.38) with $M = I_n,\quad L = 0,\quad N = D^2$. The corresponding Hamiltonian is $H = \begin{pmatrix} 0 & -I_n \\ D^2 & 0 \end{pmatrix}$. We partition D in singular and non-singular parts:

$$D = \begin{pmatrix} 0 & 0 \\ 0 & \tilde{D} \end{pmatrix}, \quad \text{where} \tilde{D} := \begin{pmatrix} \omega_1 I_2 & \dots & 0 \\ \vdots & \ddots & \vdots \\ 0 & \dots & \omega_k I_2 \end{pmatrix},$$

where $\tilde{D}$ is positive definite since all $\omega_j > 0$. Partitioning H correspondingly:

$$H = \left(\begin{array}{cc|cc} 0 & 0 & -I_{n-2k} & 0 \\ 0 & 0 & 0 & -I_{2k} \\ \hline 0 & 0 & 0 & 0 \\ 0 & \tilde{D}^2 & 0 & 0 \end{array}\right). \tag{8.41}$$

This matrix is then exponentiated to solve system (8.39):

$$\begin{pmatrix} P(t) \\ \hline Q(t) \end{pmatrix} = \left(\begin{array}{cc|cc} I_{n-2k} & 0 & -tI_{n-2k} & 0 \\ 0 & c & 0 & -\tilde{D}^{-1}s \\ \hline 0 & 0 & I_{n-2k} & 0 \\ 0 & \tilde{D}s & 0 & c \end{array}\right) \begin{pmatrix} P(0) \\ \hline Q(0) \end{pmatrix},$$

where we have introduced the following notation:

$$s(t) := \operatorname{diag}(\sin(\omega_1 t)I_2, \dots, \sin(\omega_k t)I_2) = \sin(\tilde{D}t),$$

and similarly $c(t) = \cos(\tilde{D}t)$. By setting $P(0) = I_n$, and $Q(0) = \tilde{S}_0$, and using Lemma 8.13, it follows that the solution of the initial value problem 8.37 is given by $\tilde{S}(t) = Q(t)P(t)^{-1}$,

$$\left(\frac{P(t)}{Q(t)}\right) = \left(\frac{\begin{pmatrix} (I_{n-2k} - t)\tilde{S}_0 & 0 \\ 0 & c(t) - \tilde{D}^{-1}s(t)\tilde{S}_0 \end{pmatrix}}{\begin{pmatrix} I_{n-2k}\tilde{S}_0 & 0 \\ 0 & \tilde{D}s(t) + c(t)\tilde{S}_0 \end{pmatrix}}\right), \tag{8.42}$$

for all t for which $P(t)$ is non-singular. We now make the following assumption:

Assumption A The matrix $P(t)$ is non-singular for all t in $[0, \bar{t})$, where $\bar{t}$ is finite and such that $s(t)$ is non-singular for all t in $(0, \bar{t})$. Moreover, $P(\bar{t})$ has rank $n - 1$, or equivalently, has a simple eigenvalue at zero.

Later we will show that this assumption is generically satisfied, and also that

$$t^* = \bar{t}, \tag{8.43}$$

where $[0, t^*)$ is the maximal forward interval of existence of the solution $\tilde{S}(t)$ of the initial value problem (8.37). Consequently, the theory of ODE's implies that $\lim_{t\to\bar{t}} |\tilde{S}(t)|_F = +\infty$, i.e. that $\bar{t}$ is the blow-up time for the solution $\tilde{S}(t)$.

Assuming for the moment that assumption A is satisfied, back-transformation using Eqs. (8.34) and (8.36), yields that the solution $S(t)$ of (8.31) is $S(t) = e^{tA_0} V\tilde{S}(t)V^\top e^{-tA_0}$, which is defined for all t in $[0, \bar{t})$, because $e^{tA_0} V$ is bounded for all t (as it is an orthogonal matrix). It follows from unitary invariance of the Frobenius norm that

$$\lim_{t\to\bar{t}} \frac{S(t)}{|S(t)|_F} = e^{\bar{t}A_0} V \left(\lim_{t\to\bar{t}} \frac{\tilde{S}(t)}{|\tilde{S}(t)|_F}\right) V^\top e^{-\bar{t}A_0}, \tag{8.44}$$

provided that at least one of the two limits exists. If we partition $\tilde{S}_0$ in Eq. (8.42) as follows:

$$\tilde{S}_0 = \begin{pmatrix} (\tilde{S}_0)_{11} & (\tilde{S}_0)_{12} \\ (\tilde{S}_0)_{12}^\top & (\tilde{S}_0)_{22} \end{pmatrix}, \quad \text{with} \quad \begin{matrix} (\tilde{S}_0)_{11} = (\tilde{S}_0)_{11}^\top, \\ (\tilde{S}_0)_{22} = (\tilde{S}_0)_{22}^\top, \end{matrix}$$

we can rewrite $P(t)$ and $Q(t)$ on the time interval $(0, \bar{t})$ as: $P(t) = \Delta(t)M(t)$ with,

$$\Delta(t) = \begin{pmatrix} tI_{n-2k} & 0 \\ 0 & \tilde{D}^{-1}s(t) \end{pmatrix},$$

and

$$M(t) = \begin{pmatrix} 1/t - (\tilde{S}_0)_{11} & -(\tilde{S}_0)_{12} \\ -(\tilde{S}_0)_{12}^\top & \tilde{D}c(t)s^{-1}(t) - (\tilde{S}_0)_{22} \end{pmatrix} = M^\top(t),$$

and

$$Q(t) = \begin{pmatrix} (\tilde{S}_0)_{11} & (\tilde{S}_0)_{12} \\ c(t)(\tilde{S}_0)_{12}^\top & \tilde{D}s(t) + c(t)(\tilde{S}_0)_{22} \end{pmatrix}.$$

Note that the factorization of $P(t)$ is well-defined on $(0, \bar{t})$ because by assumption A, the matrix $s(t)$ is non-singular in the interval $(0, \bar{t})$. Moreover, assumption A also implies there exists a non-zero vector u corresponding to the zero eigenvalue of $M(\bar{t})$, i.e. $M(\bar{t})u = 0$, and that u is uniquely defined up to scalar multiplication because the zero eigenvalue is simple. More explicitly, partitioning u as $\begin{pmatrix} u_1 \\ u_2 \end{pmatrix}$, there holds that

$$\begin{pmatrix} 1/\bar{t} - (\tilde{S}_0)_{11} & -(\tilde{S}_0)_{12} \\ -(\tilde{S}_0)_{12}^\top & \tilde{D}c(\bar{t})s^{-1}(\bar{t}) - (\tilde{S}_0)_{22} \end{pmatrix} \begin{pmatrix} u_1 \\ u_2 \end{pmatrix} = 0. \quad (8.45)$$

Notice that $M(t)$ is at least real-analytic on the interval $(0, \bar{t})$. Hence, it follows from [7] (see also [3, 13]), that there is an orthogonal matrix $U(t)$, and a diagonal matrix $\Lambda(t)$, both real-analytic on $(0, \bar{t})$, such that: $M(t) = U(t)\Lambda(t)U^\top(t)$, for $t \in (0, \bar{t})$, and thus $M^{-1}(t) = U(t)\Lambda^{-1}(t)U^\top(t)$, for $t \in (0, \bar{t})$. Returning to Eq. (8.44), we obtain that:

$$\begin{aligned} \lim_{t\to\bar{t}} \frac{S(t)}{|S(t)|_F} &= \mathrm{e}^{\bar{t}A_0}\, V \lim_{t\to\bar{t}} \frac{Q(t)U(t)\Lambda^{-1}(t)U^\top(t)\Delta^{-1}(t)}{|Q(t)U(t)\Lambda^{-1}(t)U^\top(t)\Delta^{-1}(t)|_F} V^\top\, \mathrm{e}^{-\bar{t}A_0} \\ &= \mathrm{e}^{\bar{t}A_0}\, V \frac{Q(\bar{t})uu^\top\Delta^{-1}(t)}{|Q(\bar{t})uu^\top\Delta^{-1}(t)|_F} V^\top\, \mathrm{e}^{-\bar{t}A_0}. \end{aligned}$$

Here, we have used the fact that $M^{-1}(t)$ is positive definite on the interval $(0, \bar{t})$, so that its largest eigenvalue (which is simple for all $t < \bar{t}$ sufficiently close to $\bar{t}$, because of assumption A approaches $+\infty$ and not $-\infty$ as $t \to \bar{t}$. To see this, note that from its definition follows that $M(t)$ is positive definite for all sufficiently small $t > 0$, because $\tilde{D}$ is positive definite. Moreover, $M(t)$ is non-singular on $(0, \bar{t})$ since by assumption (**A**), $P(t)$ is non-singular on $(0, \bar{t})$, and because $M(t) = \Delta^{-1}(t)P(t)$ (it is clear from its definition and assumption A that $\Delta(t)$ is non-singular on $(0, \bar{t})$ as well). Consequently, the smallest eigenvalue of $M(t)$ remains positive in $(0, \bar{t})$, as it approaches zero as $t \to \bar{t}$. This implies that the largest eigenvalue of $M^{-1}(t)$ is positive on $(0, \bar{t})$, and approaches $+\infty$ as $t \to \bar{t}$, as claimed.

Note that:

$$\begin{aligned} Q(\bar{t})u &= \begin{pmatrix} (\tilde{S}_0)_{11} & (\tilde{S}_0)_{12} \\ c(\bar{t})(\tilde{S}_0)_{12}^\top & \tilde{D}s(\bar{t}) + c(\bar{t})(\tilde{S}_0)_{22} \end{pmatrix} \begin{pmatrix} u_1 \\ u_2 \end{pmatrix} \\ &= \begin{pmatrix} (1/\bar{t})u_1 \\ \tilde{D}s^{-1}(\bar{t})u_2 \end{pmatrix} = \Delta^{-1}(\bar{t})u, \end{aligned}$$

where in the second equality, we used the second row of Eq. (8.45), multiplied by $c(\bar{t})$. From this follows that

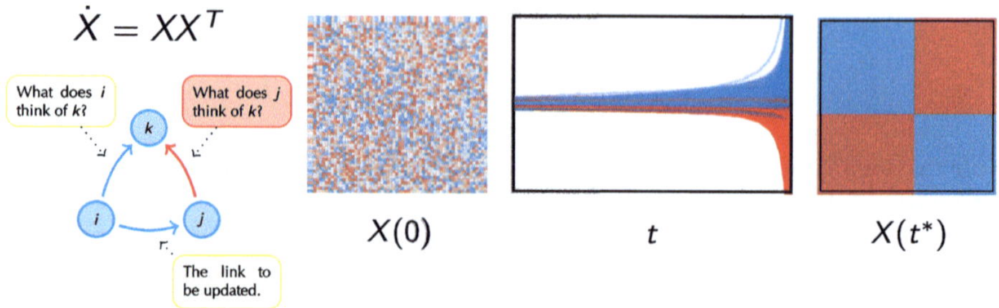

Fig. 8.3 Generic behaviour of $\dot{X} = XX^\top$

$$\lim_{t\to\bar{t}} \frac{S(t)}{|S(t)|_F} = \mathrm{e}^{\bar{t}A_0}\, V \frac{\Delta^{-1}(\bar{t})uu^\top \Delta^{-1}(\bar{t})}{|\Delta^{-1}(\bar{t})uu^\top \Delta^{-1}(\bar{t})|_F} V^\top \mathrm{e}^{-\bar{t}A_0} = \frac{ww^\top}{|ww^\top|_F},$$

where $w := e^{\bar{t}A_0} V \Delta^{-1}(\bar{t})u$.

Taking limits for $t \to \bar{t}$ in Eq. (8.33), and using the above equality, we finally arrive at the following result, which implies that system (8.30) evolves to a socially balanced state (in normalized sense) when $t \to \bar{t}$:

Proposition 8.14 *Suppose that assumption A holds and $A_0 \neq 0$. Then the solution $X(t)$ of (8.30) satisfies:*

$$\lim_{t\to\bar{t}} \frac{X(t)}{|X(t)|_F} = \frac{ww^\top}{|ww^\top|_F}$$

with $w = e^{\bar{t}A_0} V \Delta^{-1}(\bar{t})u$.

This generic behaviour is illustrated in Fig. 8.3.

8.3.3 Genericity

Generically, assumption A holds, and (8.43) holds as well. There are two aspects to assumption A:

1. The matrix $P(t)$ is non-singular in the interval $[0, \bar{t})$, but singular at some finite $\bar{t}$ such that:
$$\bar{t} < \min_{j=1,\dots,k} \frac{\pi}{\omega_j}. \tag{8.46}$$

2. $P(\bar{t})$ has a simple zero eigenvalue.

To deal with the first item, suppose that the solution $\tilde{S}(t)$ of Eq. (8.37) is defined for all $t \in [0, t^*)$ for some finite positive t^*. By Lemma 8.13, there exist $P(t)$ and $Q(t)$ such that $\tilde{S}(t) = Q(t)P^{-1}(t)$, where $P(t)$ and $Q(t)$ are components of the solution of system (8.39) with H defined in Eq. (8.41). Then necessarily $\bar{t} \leq t^*$. Thus, if we can show that $t^* < \min_j \pi/\omega_j$, then Eq. (8.46) holds. To show that $t^* < \min_j \pi/\omega_j$, we rely on a particular property of matrix Riccati differential equations (8.38): their solutions preserve the order generated by the cone of non-negative symmetric matrices (see [4]). More precisely, if $S_1(t)$ and $S_2(t)$ are solutions of Eq. (8.38), and if $S_1(0) \preceq S_2(0)$, then $S_1(t) \preceq S_2(t)$, for all $t \geq 0$ for which both solutions are defined. The partial order notation $S_1(t) \preceq S_2(t)$ means that the difference $S_2(t) - S_1(t)$ is a positive semi-definite matrix.

We apply this to Eq. (8.37) with $\tilde{S}_1(0) = \alpha_{\min} I_n$ and $\tilde{S}_2(0) = \tilde{S}(0)$, where we choose $\alpha_{\min}$ as the smallest eigenvalue of $\tilde{S}(0)$ (or equivalently, of $S(0) = S_0$, since $\tilde{S}(0) = V^\top S_0 V$), so that clearly $\tilde{S}_1(0) \preceq \tilde{S}_2(0)$. Consequently, by the monotonicity property of system (8.37), it follows that $\tilde{S}_1(t) \preceq \tilde{S}(t)$, as long as both solutions are defined. We can calculate the blow-up time t_1^* of $\tilde{S}_1(t)$ explicitly, and then it follows that $t^* \leq t_1^*$, where t^* is the blow-up time of $\tilde{S}(t)$. Indeed, equations of system (8.37) decouple for an initial condition of the form $\alpha_{\min} I_n$, and the resulting scalar equations are scalar Riccati equations we have solved before. The blow-up time for $\tilde{S}_1(t)$ is given by:

$$t_1^* = \begin{cases} \min_{j=1,\ldots,k} \left(\frac{\pi}{2\omega_j} - \frac{\phi_j}{\omega_j} \right), & \text{if } \alpha_{\min} \leq 0 \\ \min_{j=1,\ldots,k} \left(\frac{1}{\alpha_{\min}}, \frac{\pi}{2\omega_j} - \frac{\phi_j}{\omega_j} \right), & \text{if } \alpha_{\min} > 0. \end{cases}$$

with $\phi_j := \arctan\left(\frac{\alpha_{\min}}{\omega_j}\right) \in \left(-\frac{\pi}{2}, \frac{\pi}{2}\right)$. Notice that for all $j = 1, \ldots, k$, there holds that $\frac{\pi}{2\omega_j} - \frac{\phi_j}{\omega_j} < \frac{\pi}{\omega_j}$, because by definition, $\frac{\phi_j}{\omega_j} \in (-\frac{\pi}{2\omega_j}, \frac{\pi}{2\omega_j})$. Consequently,

$$\bar{t} \leq t^* \leq t_1^* < \min_{j=1,\ldots,k} \frac{\pi}{\omega_j},$$

which establishes Eq. (8.46). In other words, we have shown that the first item in assumption A is always satisfied.

The second item in assumption A may fail, but holds for generic initial conditions as we show next. For this we first point out that the derivative of each eigenvalue of $M(t)$ is a strictly decreasing function in the interval $(0, \bar{t})$, independently of the value of the matrix $\tilde{S}_0$. Indeed, the derivative of eigenvalue $\Lambda_j(t)$ of $M(t)$ equals (see [7]):

$$\dot{\Lambda}_j(t) = u_j(t)^\top \dot{M}(t) u_j(t) = -u_j(t)^\top \begin{pmatrix} 1/t^2 & 0 \\ 0 & \tilde{D}^2 s^{-2}(t) \end{pmatrix} u_j(t),$$

where $u_j(t)$ is the normalized eigenvector of $M(t)$ corresponding to $\Lambda_j(t)$, and which is analytic in the considered interval. Since $\dot{M}(t)$ is negative definite in that interval, $\dot{\Lambda}_j(t)$ is also negative and hence all eigenvalues of $M(t)$ are strictly decreasing functions of t in that interval. Suppose now that $M(t)$ has a multiple eigenvalue 0 at $t = \bar{t}$, then $M(\bar{t})$ is positive semi-definite since $\bar{t}$ is the first singular point of $M(t)$ and the eigenvalues are decreasing function of t. If we now choose a positive semi-definite $\Delta_{\tilde{S}_0}$ of nullity 1, such that $M(\bar{t}) + \Delta_{\tilde{S}_0}$ also has nullity 1, then the perturbed initial condition $(\tilde{S}_0)_p = \tilde{S}_0 - \Delta_{\tilde{S}_0}$ yields the perturbed solution $\tilde{S}_p(t)$ which can be factored as $Q_p(t)P_p^{-1}(t)$, and where $P_p(t) = \Delta(t)M_p(t)$ (note that $\Delta(t)$ remains the same as before the perturbation) for $M_p(t) = M(t) + \Delta_{\tilde{S}_0}$ which now has a single root at the same minimal value $\bar{t}$. To construct such a matrix $\Delta_{\tilde{S}_0}$ is simple since the only condition it needs to satisfy is that $M(\bar{t})$ and $\Delta_{\tilde{S}_0}$ have a common null vector. Those degrees of freedom show that the second item in assumption A is indeed generic.

Now that we have established that A generically holds, we show that Eq. (8.43) is satisfied also. The proof is by contradiction. Earlier, we have shown that $\bar{t} \leq t^*$. Thus, if we suppose that Eq. (8.43) fails, then necessarily $\bar{t} < t^*$. This implies that although $P(\bar{t})$ is singular, the solution $\tilde{S}(t)$ exists for $t = \bar{t}$. Our goal is to show that $\lim_{t\to\bar{t}} |\tilde{S}(t)|_F = +\infty$, which yields the desired contradiction (by the theory of ODE's).

We first claim the following:

$$\text{If } u \neq 0 \text{ and } P(\bar{t})u = 0, \quad \text{then } Q(\bar{t})u \neq 0. \tag{8.47}$$

Indeed, if this were not the case, then there would exist some vector $\bar{u} \neq 0$ such that $P(\bar{t})\bar{u} = 0$ and $Q(\bar{t})\bar{u} = 0$. On the other hand, $P(t)$ and $Q(t)$ are components of the matrix product

$$\begin{pmatrix} P(t) \\ Q(t) \end{pmatrix} = e^{tH} \begin{pmatrix} I_n \\ \tilde{S}_0 \end{pmatrix},$$

where H is defined in Eq. (8.41). Multiplying the latter in $t = \bar{t}$ by $\bar{u}$, and using the previous expression, it follows from the invertibility of $e^{\bar{t}H}$ that $\bar{u} = 0$, a contradiction. This establishes Eq. (8.47).

In the previous section, we factored $P(t)$ as $P(t) = \Delta(t)M(t)$. Since $P(t)$ is non-singular on $[0, \bar{t})$, and singular at $\bar{t}$, it follows from Eq. (8.46) and the definition of $\Delta(t)$, that $M(t)$ is non-singular (and, in fact, positive definite as shown in the previous section) on $(0, \bar{t})$, and singular at $\bar{t}$ as well. Therefore, since $M(t)$ is symmetric and real-analytic, it follows from [7] that we can find a positive and real-analytic scalar function $\epsilon(t)$, and a real-analytic unit vector $u(t)$ such that:

$$M(t)u(t) = \epsilon(t)u(t), \; \epsilon(t) > 0 \text{ on } (0, \bar{t}), \; \epsilon(\bar{t}) = 0, \; |u(t)|_2 = 1,$$

where $|.|_2$ denotes the Euclidean norm. In particular, $M(\bar{t})u(\bar{t}) = 0$, and since $\Delta(\bar{t})$ is non-singular, it follows that $P(\bar{t})u(\bar{t}) = 0$. Then Eq. (8.47) implies that $Q(\bar{t})u(\bar{t}) \neq 0$. Define the real-analytic unit vector

$$v(t) = \frac{\Delta(t)u(t)}{|\Delta(t)u(t)|_2}, \quad t \in (0, \bar{t}),$$

and calculate

$$\lim_{t\to\bar{t}} |\tilde{S}(t)v(t)|_2 = \lim_{t\to\bar{t}} |Q(t)P^{-1}(t)v(t)|_2$$
$$= \frac{|Q(\bar{t})u(\bar{t})|_2}{|\Delta(\bar{t})u(\bar{t})|_2} \lim_{t\to\bar{t}} \frac{1}{\epsilon(t)} = +\infty.$$

Since for any real $n \times n$ matrix A, and for any unit vector x (i.e. $|x|_2 = 1$) holds that $|Ax|_2 \leq |A|_F$, it follows that $\lim_{t\to\bar{t}} |\tilde{S}(t)|_F = +\infty$. This yields the sought-after contradiction.

By combining Proposition 8.14 and the results in this section, we have proved the main result concerning the generic emergence of balance for solutions of system Eq. (8.30).

Theorem 8.15 *There exists a dense set of initial conditions X_0 in $\mathbb{R}^{n\times n}$ such that the corresponding solution $X(t)$ of* Eq. (8.30) *satisfies:*

$$\lim_{t\to\bar{t}} \frac{X(t)}{|X(t)|_F} = \frac{ww^\top}{|ww^\top|_F}.$$

with $w = e^{\bar{t}A_0} V \Delta^{-1}(\bar{t})u$.

Proof The set of initial conditions X_0 for which $A_0 \neq 0$ and assumption A holds is dense in $\mathbb{R}^{n\times n}$. □

Summarizing, the model $\dot{X} = X^2$ does not lead to social balance generically, whereas the model $\dot{X} = XX^\top$ does. The difference between the two models can be understood in terms of gossiping: the transpose model assumes that people who wish to revise their opinion about someone talk to that person about everybody else, while the earlier model assumed that people talk about that person to everybody else. Given that social balance often holds to some degree, it is more likely that people change their opinions of each other based on the model $\dot{X} = XX^\top$ then on $\dot{X} = X^2$.

References

1. Abou-Kandil H (2003) Matrix Riccati equations: in control and systems theory. Springer. ISBN 376430085X
2. Antal T, Krapivsky P, Redner S (2005) Dynamics of social balance on networks. Phys Rev E 72(3):36121. doi:10.1103/PhysRevE.72.036121
3. Bunse-Gerstner A, Byers R, Mehrmann V, Nichols NK (1991) Numerical computation of an analytic singular value decomposition of a matrix valued function. Numerische Mathematik 60(1):1–39. doi:10.1007/BF01385712
4. De Leenheer P, Sontag ED (2004) A note on the monotonicity of matrix Riccati equations. Technical report, DIMACS
5. Durrett R, Levin SA (2005) Can stable social groups be maintained by homophilous imitation alone? J Econ Behav Organ 57(3):267–286. doi:10.1016/j.jebo.2003.09.017

6. Fu F, Nowak MA, Christakis NA, Fowler JH (2012) The evolution of homophily. Sci Rep 2. doi:10.1038/srep00845
7. Kato T (1995) Perturbation theory for linear operators. Springer, New York. ISBN 354058661X
8. Kulakowski K, Gawronski P, Gronek P (2005) The Heider balance–a continuous approach. Int J Mod Phys C 16(5):707–716. doi:10.1142/S012918310500742X
9. Marvel S, Strogatz S, Kleinberg J (2009) Energy landscape of social balance. Phys Rev Lett 103(19):198701. doi:10.1103/PhysRevLett.103.198701. arXiv:0906.2893
10. Marvel SA, Kleinberg J, Kleinberg RD, Strogatz SH (2011) Continuous-time model of structural balance. Proc Nat Acad Sci USA 108(5):1771–6. doi:10.1073/pnas.1013213108
11. Mcpherson M, Smith-Lovin L, Cook JM (2001) Birds of a feather: homophily in social networks. Annu Rev Soc 27(1):415–444. doi:10.1146/annurev.soc.27.1.415
12. Rugh W (1996) Linear system theory. Prentice Hall, Upper Saddle River. ISBN 9780134412054
13. Still G (2001) How to split the eigenvalues of a one-parameter family of matrices. Optimization 49(4):387–403. doi:10.1080/02331930108844539

Chapter 9
Evolution of Cooperation

We will now discuss a subject that perhaps seems remote: the evolution of cooperation. For an introduction to the field, Nowak [20] provides a light read, while Hofbauer and Sigmund [12] provide a more technical background. Although the subject is not commonly considered when talking about negative links, the two are naturally related. A positive link indicates people are cooperating, while a negative link indicates people are not cooperating with each other. We will first introduce the general subject, which is usually studied through the so-called prisoner's dilemma or variants thereof. As we will see, it is usually better to defect (i.e. not cooperate), and evolution tends to favour defection. The biggest question in this subject therefore is: why do we so frequently then observe cooperation? In other words, under what conditions can we expect cooperative behaviour to evolve? We will first briefly discuss two types of answers to this question. The first concerns repeated interaction and is known as "direct reciprocity" [3]. The latter concerns reputation and the transfer of information, known as "indirect reciprocity" [25]. Finally, we will consider what this has to do with negative links, and how this involves the previously discussed models of social balance.

9.1 Game Theory

The history of the evolution of cooperation is a long and tumultuous one, and cooperation already posed problems for Darwin: if defection allows individuals to obtain a higher fitness, then why do we observe cooperation? Although Darwin himself did try to explain the situation, especially concerning the cooperative behaviour of insects, it was not until around the second world war that the problem was formalised.

In 1944 Von Neumann and Morgenstern published the "Theory of Games and Economic behaviour" which was the first push towards the formalisation of the evolution of cooperation [37]. They studied various games, in which parties could take independent decisions, each with a different payoff, and the focus was on determining the optimal decision (those which maximised the payoff). The famous contribution

V. Traag, *Algorithms and Dynamical Models for Communities and Reputation in Social Networks*, Springer Theses, DOI: 10.1007/978-3-319-06391-1_9,

of John Nash was that he proved that for each game (with a finite set of decision) such an optimal strategy exists [18]. Such an optimal strategy means that knowing the opponents strategy, you choose the "best response". If both strategies are the best response to each other, such a pair is called a Nash equilibrium nowadays in his honour.

Another type of optimality condition however, is that of a Pareto optimum. In such an "optimal condition" no player can increase his payoff without making another player worse off. The two conditions of optimality do not necessarily agree however, which then might results in a paradox. Although no player would individually choose a different strategy, every player might be better off with an alternative strategy.

One of these specific games has become quite famous over the years, and is known as the prisoner's dilemma. Originally, the story goes as follows. Two suspects have been apprehended by the police after a robbery, say Jack and Harry. However, the police has some difficulty in proving the involvement of the two suspects, and if both prisoners don't talk, they can only convict them of a minor crime with a sentence of only 1 year. However, the police tries to get a more severe conviction, and if one of the prisoners rats out the other one, they will be able to get a conviction of 5 years for the other one. The police decide to proposes the following deal to both prisoners: "If you talk, we will reduce your sentence by 1 year." Hence, if Jack betrays Harry while Harry doesn't, Jack will get off with no jail time, while Harry will have to do 5 years in prison. On the other hand, if Jack also betrays Harry, then Harry can better betray Jack as well, since he will then only serve 4 years instead of 5. So, in this case, for both Harry and Jack individually it is better to betray each other (betrayal is a Nash equilibrium). On the other hand, it is obvious that the best situation for both is that they do not betray each other (remaining silent is a Pareto optimum).

Although these conditions of optimality provide an interesting qualification of the quality of a certain strategy, the focus here is on the evolution of a certain behaviour [15]. The study in this context was first put forward by Maynard Smith and Price [16], although a similar concept was also considered in another form by Hamilton [10]. Their concept of an Evolutionary Stable Strategy (ESS) is actually very similar to that of a Nash equilibrium. Furthermore, it is natural to specify some evolutionary dynamics, in order to state whether some specific strategy is evolutionary stable: these are simply the fixed points of the evolutionary dynamics.

We will now formalize these issues here, and briefly review some of the concepts. We will restrict ourselves to symmetrical two player games. Furthermore, there is a distinction between a finite population and an infinite population. The first is usually modelled using discrete time and are inherently stochastic in nature, while the latter gives rise to deterministic differential equations. However, the differential equations arise as a limit of large population size for some of the finite population models.

First, we define the set $S = \{1, \dots, q\}$ of the different possible options. Player 1 and 2 may each choose an option $i \in S$ and $j \in S$. The payoff for player 1 is then A_{ij} while for player 2 the payoff is A_{ji}. We allow both players to choose option i with a probability x_i, and call this vector x the strategy of the player, for which $\sum x_i = 1$. We call x a pure strategy if $x_i = 1$ for some i (so that only choice i is used), and a mixed strategy otherwise [12].

Definition 9.1 Let x be a strategy and A the payoff matrix. Then x is called a Nash equilibrium if

$$y^\top Ax \leq x^\top Ax$$

for all $y \neq x$. If this inequality is strict, we call x a strict Nash equilibrium.

The idea behind this definition is that no player using strategy x will have an incentive to change to any other strategy y. This can be seen as follows. The payoff when option i is chosen versus option j is A_{ij}. Since option i is chosen with probability y_i by strategy y and option j is chosen with probability x_j by strategy x, the sum $\sum_{ij} y_i A_{ij} x_j = y^\top Ax$ is the expected payoff when a strategy y plays against a strategy x. Hence, if both players use strategy x, neither player can improve its payoff by switching to a strategy $y \neq x$. In this sense, x is said to be a best reply to itself, since if somebody plays strategy x, one should also play strategy x.

As we said earlier, the related concept of an Evolutionary Stable Strategy (ESS) was developed by biologists relatively independently of the Nash equilibrium (and some 20 years later). This can be formalized in a similar way as the Nash equilibrium.

Definition 9.2 Let A be the payoff matrix. A strategy x is called an evolutionary stable strategy (ESS) if for some $y \neq x$

1. $y^\top Ax \leq x^\top Ax$, or
2. if $y^\top Ax = x^\top Ax$ then $y^\top Ay < x^\top Ay$.

The motivation for ESS is the following. Let x and y be two strategies. Suppose that we consider a third strategy that consists of the convex combination of the two strategies $z = \varepsilon x + (1 - \varepsilon)y$. The idea is then that moving a bit from strategy x in the direction towards y should not increase the payoff. Hence $y^\top Az < x^\top Az$ which after rewriting gives

$$(1 - \varepsilon)(x^\top Ax - y^\top Ax) + \varepsilon(x^\top Ay - y^\top Ay) > 0$$

which yields the stated inequalities. The first condition simply states that the strategy should be a Nash equilibrium, while the second condition states that if another strategy y is equally well against x, then it should be less well against itself than x against y. Obviously, if a strategy is a strict Nash equilibrium, it is ESS, which in turn implies that the strategy is a (weak) Nash equilibrium.

For our small example of the prisoner's dilemma, we thus have the following payoff matrix

$$A = -\begin{array}{c} \\ C \\ D \end{array}\!\!\begin{array}{c} \begin{array}{cc} C & D \end{array} \\ \begin{pmatrix} 1 & 5 \\ 0 & 4 \end{pmatrix} \end{array},$$

where cooperation (denoted by C) here means to keep silent, and defection (denoted by D) means to betray the other and talk to the police. The payoffs here are negative since they correspond to years in jail, something most people presumably like to

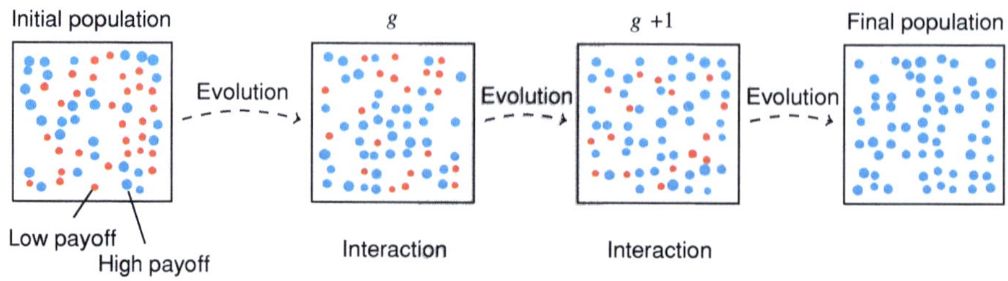

Fig. 9.1 Schematic of the evolutionary dynamics

avoid. Defecting corresponds to the pure strategy $x = (0, 1)$, with payoff $x^\top Ax = -4$. For any other strategy $y = (p, 1 - p)$ we obtain

$$y^\top Ax = -5p - (1-p)4 < -4 = x^\top Ax,$$

so that indeed defecting is a strict Nash equilibrium, hence also an evolutionary stable strategy. Nonetheless, the strategy $y = (1, 0)$ has payoff

$$y^\top Ay = -1 > -4 = x^\top Ax,$$

and so y is preferable for both players.

9.1.1 Finite Population Size

The basic idea of evolution is that of reproduction and fitness dependent selection. Hence, agents that have a higher fitness, and are more likely to reproduce (or less likely to die), as illustrated in Fig. 9.1. There are various possible scenarios here, and we will discuss only a few of them, most notably the Moran process [22], the pairwise comparison [35] (sometimes also known as the Fermi process) and the Wright-Fisher process [13].

Let us assume there are n agents, and that each agent is of some type $T_i \in S$. We denote by $n_s = |\{T_i = s\}|$ the number of agents that are of type s, and of course $\sum_s n_s = n$. We denote the vector of n_s by $\vec{n}$. We assume each agent is interacting randomly with some other agents, and that no specific population structure is present. We therefore assume that the fitness for each agent of the same type is equal, and we denote the fitness for an agent of type s by $f_s(\vec{n})$, which may depend on the number of agents of some specific type.

Moran Process

The Moran Process is very simple: we select an agent for reproduction with some probability proportional to their fitness [22]. So, the probability that some agent of type s reproduces is given by

$$\phi_s = \frac{n_s f_s(\vec{n})}{n\langle f\rangle} \tag{9.1}$$

where $\langle f\rangle = \frac{1}{n}\sum_s n_s f_s$ is the average fitness of the population. We then randomly choose another agent to die, which will be replaced by the offspring from the agent chosen for reproduction. This probability is denoted by $\omega_s = \frac{n_s}{n}$. We denote accordingly by $n_s(g)$ the number of agents of type s in generation g. The difference between two generations can then be denoted by $\Delta n_s(g) = n_s(g+1) - n_s(g)$. The update rules introduced above can then be written explicitly as

$$\Pr(\Delta n_s(g) = -1) = (1-\phi_s)\omega_s, \tag{9.2a}$$

$$\Pr(\Delta n_s(g) = 0) = \phi_s\omega_s + (1-\phi_s)(1-\omega_s), \tag{9.2b}$$

$$\Pr(\Delta n_s(g) = 1) = \phi_s(1-\omega_s). \tag{9.2c}$$

For randomly mixing populations, the order of reproduction and dying has no effect, but for structured populations it does. We will not discuss this here.

Pairwise Comparison

Another possibility instead of selecting one agent for reproduction and another one for death, is to compare the relative fitness between two agents, and let them compete to "take over" the other one's spot, so to speak. One of the advantages is that we don't need to know the fitness of all agents, but only of the two agents [34, 35]. These dynamics are known as the pairwise comparison process (also the Fermi process). If we compare an agent of type r to type s, the probability that an agent of type r replaces an agent of type s is then given by

$$\phi_{rs} = \frac{1}{1+e^{-\beta(f_r(\vec{n})-f_s(\vec{n}))}}. \tag{9.3}$$

The parameter β corresponds to a certain intensity of selection. For high β it is almost sure that if r has a higher fitness than s that he will "win" the competition and $\phi_{rs} \to 1$ for $\beta \to \infty$ if $f_r(\vec{n}) > f_s(\vec{n})$. For low β on the other hand, almost all types will have the same probability to reproduce, independently of their fitness, and that $\phi_{rs} \to 1/2$ for $\beta \to 0$. Notice that in general $\phi_{sr} = 1 - \phi_{rs}$. The number of agents of a certain type can only be augmented by 1, and we obtain the following expressions

$$\Pr(\Delta n_s(g) = -1) = \frac{n_s(g)}{n}\sum_{r\neq s}\frac{n_r(g)}{n}\phi_{rs}, \tag{9.4a}$$

$$\Pr(\Delta n_s(g) = 0) = \frac{n_s^2(g)}{n}, \tag{9.4b}$$

$$\Pr(\Delta n_s(g) = 1) = \frac{n_s(g)}{n}\sum_{r\neq s}\frac{n_r(g)}{n}\phi_{sr}. \tag{9.4c}$$

Wright-Fisher Process

Whereas the previous models only modelled the reproduction and death of a single agent, in the Wright-Fisher process the whole population evolves simultaneously [13]. In this process one "evolutionary step" consists of randomly sampling n agents from the old population based on their fitness, to form the new population. Hence

$$\phi_s = \frac{n_s f_s(\vec{n})}{n\langle f\rangle} \tag{9.5}$$

similar to the Moran process. However, now instead of selecting a single agent, we select n agents. Hence, the probability that there are n_s agents of type s in the next generation is

$$\Pr(n_s(g+1) = n_s) = \binom{n_s}{n} \phi_s^{n_s}(1-\phi_s)^{n-n_s} \tag{9.6}$$

and the expected number of agents of type s is $\langle n_s(g+1)\rangle = n\phi_s$. Hence, a higher fitness $f_s(\vec{n})$ leads to a higher ϕ_s, which in turn leads to a higher expected number of agents in the next generation.

9.1.2 Fixation Probability for 2 × 2 Games

Let us now focus on 2×2 games, where we assume there are two types of players, A players and B players. This corresponds to a payoff matrix with

$$A = \begin{matrix} & \begin{matrix} A & B \end{matrix} \\ \begin{matrix} A \\ B \end{matrix} & \begin{pmatrix} a & b \\ c & d \end{pmatrix} \end{matrix}. \tag{9.7}$$

Strategy A, corresponding to the pure strategy $x = (1, 0)$ is then a Nash equilibrium if $a \geq c$ (and strictly Nash if the inequality is strict) and similarly strategy B is a Nash equilibrium if $d \geq b$. Strategy A is an ESS if $a > c$ or if $a = c$ then $b > d$.

We are interested in the probability that A players (or B players) will take over the whole population starting from n_0 A players (or B players). This quantity is known as the fixation probability $\rho_A(n_0)$ (or $\rho_B(n_0)$ for B players) [22]. We will focus on the Moran process and the pairwise comparison, which behave rather similar. It can be seen that this amounts to a sort of biased random walk on the number of A players (or B players). Let us denote the number of A players by i, and denote the transition probabilities as follows

$$\Pr(\Delta n_A = -1|i) = \lambda_i^- \tag{9.8a}$$

$$\Pr(\Delta n_A = 0|i) = \lambda_i^0 \tag{9.8b}$$

$$\Pr(\Delta n_A = +1|i) = \lambda_i^+. \tag{9.8c}$$

Clearly $\lambda_i^- + \lambda_i^0 + \lambda_i^+ = 1$, and so we can also write $\lambda_i^0 = 1 - \lambda_i^- - \lambda_i^+$. Obviously, $\lambda_0^0 = \lambda_n^0 = 1$ and the states $i = 0$ and $i = n$ are absorbing. We are then interested in the probability to reach state n (all A players) starting from state $i = n_0$, which we will denote by $\rho_i = \rho_A(i)$. The probability ρ_i to reach state n from state i is the probability ρ_{i-1} to reach it from state $i-1$ times the probability to go from state i to state $i-1$, etc., and we arrive at the recursion

$$\rho_i = \lambda_i^- \rho_{i-1} + \lambda_i^0 \rho_i + \lambda_i^+ \rho_{i+1}, \tag{9.9}$$

In addition, we have that $\rho_0 = 0$, because it is an absorbing state, and we can never reach state n from $i = 0$ and that $\rho_n = 1$ since we have already reached it. Let us introduce a new variable $y_i = \rho_i - \rho_{i-1}$. Then

$$\begin{aligned} y_{i+1} &= \rho_{i+1} - \rho_i \\ &= \rho_{i+1} - \lambda_i^- \rho_{i-1} - (1 - \lambda_i^- - \lambda_i^+)\rho_i - \lambda_i^+ \rho_{i+1} \\ &= (1 - \lambda_i^+)(\rho_{i+1} - \rho_i) + \lambda_i^- (\rho_i - \rho_{i-1}) \\ &= (1 - \lambda_i^+) y_{i+1} + \lambda_i^- y_i, \end{aligned}$$

and hence $y_{i+1} = \frac{\lambda_i^-}{\lambda_i^+} y_i$. Since $\rho_0 = 0$ we have $y_1 = \rho_1$ and we obtain

$$y_i = \rho_1 \prod_{k=1}^{i} \frac{\lambda_k^-}{\lambda_k^+}.$$

Furthermore, $\sum_{i=1}^{j} y_i = \rho_j - \rho_0 = \rho_j$, and specifically, $\sum_i y_i = \rho_n - \rho_0 = 1$ so that

$$\rho_1 = \frac{1}{1 + \sum_{j=1}^{n-1} \prod_{k=1}^{j} \frac{\lambda_k^-}{\lambda_k^+}}.$$

Moreover,

$$\begin{aligned} \rho_i &= \rho_1 \left(1 + \sum_{j=1}^{i-1} \prod_{k=1}^{j} \frac{\lambda_k^-}{\lambda_k^+} \right) \\ &= \frac{1 + \sum_{j=1}^{i-1} \prod_{k=1}^{j} \frac{\lambda_k^-}{\lambda_k^+}}{1 + \sum_{j=1}^{n-1} \prod_{k=1}^{j} \frac{\lambda_k^-}{\lambda_k^+}}. \end{aligned} \tag{9.10}$$

we thus obtain

$$\rho_A = \rho_1 = \frac{1}{1 + \sum_{i=1}^{n-1} \prod_{k=1}^{i} \frac{\lambda_k^-}{\lambda_k^+}} \tag{9.11}$$

Moreover, the ratio between $\rho_A = \rho_1$ and $\rho_B = 1 - \rho_{n-1}$ becomes simply

$$\frac{\rho_A}{\rho_B} = \prod_{i=1}^{n-1} \frac{\lambda_i^-}{\lambda_i^+} \tag{9.12}$$

If this fraction is larger then 1 and $\rho_A > \rho_B$ then type A is said to be *risk dominant*.

If $\lambda_k^- = \lambda_k^+$ for all k, then the fixation probabilities simplify to $\rho_i = \frac{i}{n}$. Hence, if the fitness for both species is always equal, the fixation probability is simply $\rho_i = \frac{i}{n}$, and we refer to this as the *neutral fixation probability*. If a species has a higher fixation probability than neutral it has an *evolutionary advantage* and if it is lower an evolutionary disadvantage.

In general, whether fixation probabilities are larger then neutral, or whether a strategy is risk dominant does not depend on the intensity of selection. We denote by

$$\mathcal{F}_i(\beta) = \prod_{k=1}^{i} \frac{\lambda_k^-}{\lambda_k^+}$$

where β represent the intensity of selection. If the intensity of selection $\beta = 0$, we expect any evolutionary process to be neutral (since otherwise there would effectively be selection). Hence, we obtain that for $\beta = 0$ that

$$\rho_A = \frac{1}{n} = \frac{1}{1 + \sum_i \mathcal{F}_i(\beta)}$$

so that $1 + \sum_i \mathcal{F}_i(\beta) = n$. Then for small β using a Taylor expansion to be evolutionary advantageous $\rho_A > 1/n$ comes down to

$$n < 1 + \sum_i \mathcal{F}_i(\beta) \approx 1 + \sum_i \mathcal{F}_i(0) + \beta \sum_i \mathcal{F}_i'(0) = n + \beta \sum_i \mathcal{F}_i'(0)$$

so that effectively if $\sum_i \mathcal{F}_i'(0) < 0$ then $\rho_A > 1/n$. Hence, to be evolutionary advantageous is independent of the intensity of selection. Similarly for risk dominance, we obtain that $\mathcal{F}_{n-1}(0) = 1$ so that strategy A is risk dominant if $\mathcal{F}'(0) > 0$ for small β. Hence, in general, these properties do not depend on the intensity of selection.

This neutral fixation probability is also valid for the Wright-Fisher process. However, the analysis of the fixation probability is more complex for the Wright-Fisher process, and we will not include it here [13, 31].

Moran Process

For the Moran process it is customary [22] to specify the fitness as

$$f_A(\vec{n}) = 1 - \beta + \beta F_A(\vec{n}), \tag{9.13a}$$

$$f_B(\vec{n}) = 1 - \beta + \beta F_B(\vec{n}), \tag{9.13b}$$

where β represents the intensity of selection and $F_A(\vec{n})$ and $F_B(\vec{n})$ are the payoffs for type A and B, which for the matrix given in Eq. (9.7) gives

$$F_A(\vec{n}) = \frac{n_A - 1}{n-1}a + \frac{n_B}{n-1}b,$$

$$F_B(\vec{n}) = \frac{n_A}{n-1}c + \frac{n_B - 1}{n-1}d.$$

The transition probabilities $\lambda_i^{\pm}$ are then

$$\lambda_i^- = \frac{n_B f_B(\vec{n})}{n\langle f\rangle}\frac{n_A}{n} = \frac{(n-i)f_B(i)}{if_A(i) + (n-i)f_B(i)}\frac{i}{n}$$

$$\lambda_i^+ = \frac{n_A f_A(\vec{n})}{n\langle f\rangle}\frac{n_B}{n} = \frac{if_A(i)}{if_A(i) + (n-i)f_B(i)}\frac{n-i}{n}$$

and the ratio becomes

$$\frac{\lambda_i^-}{\lambda_i^+} = \frac{f_B(i)}{f_A(i)}. \tag{9.14}$$

Now for general β it is difficult to analyse the product $\prod_{i=1}^{n-1} \frac{\lambda_i^-}{\lambda_i^+}$. In the limit of "weak selection" however, i.e. as $\beta \to 0$, the analysis becomes tractable. In that case, we obtain by a simple Taylor approximation

$$\frac{\lambda_i^-}{\lambda_i^+} \approx 1 + \beta\frac{\partial\frac{f_B(i)}{f_A(i)}}{\partial\beta} = 1 - \beta(F_A(i) - F_B(i)).$$

By ignoring contributions β^2 (because $\beta \to 0$) in the product, we obtain

$$\prod_{i=1}^{n-1}\frac{\lambda_i^-}{\lambda_i^+} \approx 1 - \beta\sum_{i=1}^{n-1}(F_A(i) - F_B(i)). \tag{9.15}$$

The difference $F_A(i) - F_B(i)$ can be written as

$$F_A(i) - F_B(i) = \frac{1}{n-1}(\alpha i + \omega) \tag{9.16}$$

where $\alpha = (a-b) - (c-d)$ and $\omega = n(b-d) + d - a$. In that case, the sum in Eq. (9.15) is relatively straightforward to calculate, and we obtain

$$\begin{aligned} \sum_{i=1}^{k}(F_A(i) - F_B(i)) &= \omega\frac{k}{n-1} + \frac{1}{n-1}\sum_{i=1}^{k}\alpha i \\ &= \omega\frac{k}{n-1} + \alpha\frac{k(k+1)}{2(n-1)} \end{aligned} \tag{9.17}$$

The ratio between the fixation probabilities is then

$$\frac{\rho_A}{\rho_B} = \prod_{i=1}^{n-1}\frac{\lambda_i^-}{\lambda_i^+} \approx= 1 - \beta(\omega + \alpha\frac{n}{2}). \tag{9.18}$$

Hence, if $\omega + \alpha\frac{n}{2} > 0$ then a single A mutant has a higher probability of fixation as a single B mutant, and type A is risk dominant, which is the case if

$$a(n-2) + bn > d(n-2) + cn \tag{9.19}$$

which for large n becomes approximately $a + b > d + c$. The fixation probability ρ_A itself is, from Eq. (9.11),

$$\rho_A = \frac{1}{1 + \sum_{k=1}^{n-1}\prod_{i=1}^{k}\frac{\lambda_i^-}{\lambda_i^+}}. \tag{9.20}$$

The product is

$$\prod_{i=1}^{k}\frac{\lambda_i^-}{\lambda_i^+} \approx 1 - \beta\left(\omega\frac{k}{n-1} + \alpha\frac{(k+1)k}{2(n-1)}\right) \tag{9.21}$$

and so the sum is

$$\begin{aligned} \sum_{k=1}^{n-1}\prod_{i=1}^{k}\frac{\lambda_i^-}{\lambda_i^+} &\approx \sum_{k=1}^{n-1}\left[1 - \beta\left(\omega\frac{k}{n-1} + \alpha\frac{(k+1)k}{2(n-1)}\right)\right] \\ &= (n-1) - \frac{n(n-1)\beta\omega}{2(n-1)} - \frac{\beta\alpha}{2(n-1)}\frac{(n-1)n(n+1)}{3} \\ &= (n-1) - \frac{n\beta\omega}{2} - \beta\alpha\frac{n(n+1)}{6}. \end{aligned}$$

The fixation probability thus becomes

$$\rho_A \approx \frac{1}{1+(n-1)-\frac{n\beta\omega}{2}-\frac{\beta\alpha n(n+1)}{6}} \tag{9.22}$$

$$= \frac{1}{n}\frac{1}{1-\frac{\beta\omega}{2}-\frac{\beta\alpha(n+1)}{6}} \tag{9.23}$$

Now we are interested in knowing when the fixation probability $\rho_A > 1/n$ is greater than neutral. This leads to $3\omega+\alpha(n+1) > 0$, and so whether the fixation probability is larger than neutral is independent of the intensity of selection. For large n this amounts to

$$2b + a > 2d + c. \tag{9.24}$$

Pairwise Comparison

The ratio between the transition probabilities $\lambda_i^\pm$ takes a much simpler, more elegant form for the pairwise comparison. From Eq. (9.3) we obtain

$$\lambda_i^- = \frac{i}{n}\frac{n-i}{n}\frac{1}{1+\exp(-\beta(f_A - f_B))}$$

$$\lambda_i^+ = \frac{i}{n}\frac{n-i}{n}\frac{1}{1+\exp(\beta(f_A - f_B))}.$$

The ratio between the transition probabilities therefore becomes

$$\frac{\lambda_i^-}{\lambda_i^+} = e^{-\beta(f_A - f_B)}. \tag{9.25}$$

For the pairwise comparison, we usually take as fitness simply the payoff, so that $f_A = F_A$ and $f_B = F_B$. The product of the ratio's is also quite simple

$$\prod_{i=1}^{k}\frac{\lambda_i^-}{\lambda_i^+} = \exp\left(-\beta\sum_{i=1}^{k}(f_A - f_B)\right), \tag{9.26}$$

and this holds not only for weak selection, but for all intensities of selection. We have already calculated this sum earlier in Eq. (9.17), and so we obtain

$$\prod_{i=1}^{k}\frac{\lambda_i^-}{\lambda_i^+} = \exp\left(-\beta\omega\frac{k}{n-1} + \alpha\frac{k(k+1)}{2(n-1)}\right).$$

Notice that for weak selection $\beta \to 0$ we obtain the exact same results since $e^{-\beta\Delta F} \approx 1 - \beta\Delta F$. The ratio between ρ_A and ρ_B is therefore

$$\frac{\rho_A}{\rho_B} = \exp\left[-\beta\left(\omega + \alpha\frac{n}{2}\right)\right],$$

and type A is risk dominant whenever $\omega + \alpha\frac{n}{2} > 0$, resulting in the same inequality as for the Moran process in Eq. (9.19). However, the result is valid for all intensities of selection for the pairwise comparison, and whether a strategy is risk dominant is independent of the intensity of selection.

The actual fixation probability ρ_A is

$$\rho_A = \frac{1}{1 + \sum_{k=1}^{n-1} \exp\left(-\beta\omega\frac{k}{n-1} + \alpha\frac{k(k+1)}{2(n-1)}\right)}. \tag{9.27}$$

For large n we can approximate the sum by an integral, while for weak selection $\beta \to 0$ we arrive at the same result as for the Moran process.

Notice that for the Moran process, we can also take as fitness

$$f_A(\vec{n}) = e^{\beta F_A(\vec{n})}, \tag{9.28a}$$

$$f_B(\vec{n}) = e^{\beta F_B(\vec{n})}, \tag{9.28b}$$

with β again the intensity of selection. In that case the transition probabilities $\lambda_i^{\pm}$ are

$$\lambda_i^- = \frac{(n-i)e^{\beta F_A(\vec{n})}}{(n-i)e^{\beta F_A(\vec{n})} + ie^{\beta F_B(\vec{n})}} \frac{i}{n},$$

$$\lambda_i^+ = \frac{ie^{\beta F_A(\vec{n})}}{(n-i)e^{\beta F_A(\vec{n})} + ie^{\beta F_B(\vec{n})}} \frac{n-i}{n}.$$

Notice that although this is different to the pairwise comparison process, the ratio between the transition probabilities is also

$$\frac{\lambda_i^-}{\lambda_i^+} = e^{-\beta(F_A(\vec{n}) - F_B(\vec{n}))}. \tag{9.29}$$

Hence, using Eq. (9.28) instead of Eq. (9.13) as fitness, the Moran process and the pairwise comparison process are equivalent [36] as far as the fixation probability is concerned. This is surprising since we needed information about all agents for the Moran process, while for the pairwise comparison, we only used local information, yet the two behave similarly.

9.1.3 Infinite Population Size

When the population size goes to infinity, the changes are essentially continuous, and we denote by $x_s = \frac{n_s}{n}$ the relatively frequency of type s and by x the vector of $x = (x_1, x_2, \ldots, x_q)$. The fitness of type s is $f_s(x)$ similar as before, while the average fitness is $\langle f \rangle$. Let us assume that all agents reproduces at some same basic rate α and die at some rate β, and that these rates are proportional to the number of agents (corresponding to basic exponential growth). In addition, the reproduction of some type s is based on its fitness $f_s(x)$. Furthermore, let us assume that the rate of reproduction is proportional to some time interval Δt. Then the number of agents at time $t + \Delta t$

$$n_s(t + \Delta t) = \Delta t(\alpha + f_s(x) - \beta)n_s(t) + n_s(t),$$

and so by taking the limit $\Delta t \to 0$ we obtain that

$$\lim_{\Delta t \to 0} \frac{n_s(t + \Delta t) - n_s(t)}{\Delta t} = \dot{n}_s = (\alpha + f_s(x) - \beta)n_s. \tag{9.30}$$

Since $n_s(t) = x_s(t)n(t)$, we have that $\dot{n}_s = \dot{x}_s n + x_s \dot{n}$ so that

$$\dot{x}_s n = \dot{n}_s - x_s \dot{n}. \tag{9.31}$$

Now since $n = \sum_s n_s$ we simply have that

$$\begin{aligned}\dot{n} &= \sum_s \dot{n}_s \\ &= \sum_s (\alpha + f_s(x) - \beta)n_s \\ &= (\alpha + \langle f \rangle - \beta)n.\end{aligned} \tag{9.32}$$

Then plugging Eqs. (9.32) and (9.30) into Eq. (9.31) we obtain

$$\dot{x}_s n = (\alpha - \beta)(n_s - x_s n) + f_s(x)n_s - \langle f \rangle x_s n$$

so dividing by n we arrive at

$$\dot{x}_s = (f_s(x) - \langle f \rangle)x_s. \tag{9.33}$$

This equation is known as the replicator equation [12].

In the case of game theory, we assume that each type s will play a certain (mixed) strategy p_s. The average strategy is then $p(x) = \sum_s x_s p_s$, so that the average payoff for type s is $f_s(x) = p_s^\top Ap$, while the average payoff is then $\langle f \rangle = p^\top Ap$, so that the replicator equation becomes

$$\dot{x}_s = (p_s^\top Ap(x) - p^\top(x)Ap(x))x_s. \tag{9.34}$$

If we assume each type corresponds to a pure strategy, this simplifies further. For simplicity, we can assume that A is then a $q \times q$ matrix, and that each choice coincides with a type. Hence, the strategy $p_s = e_s$ (where $e_s = \delta_{si}$ is the standard basis vector), and a type s always plays choice s. In that case, the average strategy is simply $p(x) = x$. Hence, we then arrive at

$$\dot{x}_s = (e_s^\top Ax - x^\top Ax)x_s. \tag{9.35}$$

Now there exists a certain correspondence between Nash equilibria, evolutionary stable strategies (ESS) and the stability of rest points of this replicator equation. In this case the population state x is used as the mixed strategy in the definition of a Nash equilibrium. Then the following can be proven [12].

Theorem 9.3 *In the following, statements of dynamical systems refer to Eq.* (9.35).

1. *If x is Nash, then x is a fixed point.*
2. *If x is strictly Nash, then x is asymptotically stable.*
3. *If x is a stable fixed point, then x is Nash.*
4. *If x is ESS, then x is asymptotically stable.*
5. *If x is ESS and in interior, then x is globally attracting.*

We can also explicitly analyse how the finite population size models behave for large n to see how the finite and infinite population models are connected [34, 35]. In order to develop this derivation, it is easiest to take the Moran process and pairwise comparison for the 2×2 case, so that we again only have A and B players. In that case the replicator equation reduces to

$$\dot{x} = x(1 - x)(f_A(x) - f_B(x)), \tag{9.36}$$

where $x = x_A$ and $1 - x = x_B$.

Let us denote by $P(i, \tau)$ the probability there are i A players at time τ, which then respects the recursion

$$\begin{aligned} P(i, \tau + 1) = P(i, \tau) + P(i - 1, \tau)\lambda_{i-1}^+ \\ + P(i + 1, \tau)\lambda_{i+1}^+ - P(i, \tau)(\lambda_i^+ + \lambda_i^-). \end{aligned} \tag{9.37}$$

We will now see how this probability distribution $P(i, \tau)$ changes for large n. In order to do so, we need to introduce a few equivalents. We rescale time as $t = \frac{\tau}{n}$ and denote by $x = \frac{i}{n}$ the fraction of A players. The probability there are i A players then becomes $nP(i, \tau) = \rho(x, t)$, while the transition probabilities become simply $\lambda_i^\pm = \lambda^\pm(x)$. From Eq. (9.37) we then obtain

$$\begin{aligned} \rho(x, t + 1/n) - \rho(x, t) = \rho(x - 1/n, t)\lambda^+(x - 1/n) \\ + \rho(x + 1/n, t)\lambda^-(x + 1/n) \\ - \rho(x, t)(\lambda^+(x) + \lambda^-(x)). \end{aligned} \tag{9.38}$$

For approximating the quantities $\rho(x, t + 1/n)$, $\rho(x \pm 1/n, t)$ and $\lambda^\mp(x \pm 1/n)$ we will use a Taylor expansion. We then arrive at

$$\rho(x, t + 1/n) = \rho(x, t) + \frac{1}{n}\frac{\partial \rho(x, t)}{\partial t}$$
$$\rho(x \pm 1/n, t) = \rho(x, t) \pm \frac{1}{n}\frac{\partial \rho(x, t)}{\partial x} + \frac{1}{2n^2}\frac{\partial^2 \rho(x, t)}{\partial x^2}$$
$$\lambda^{\mp}(x \pm 1/n) = \rho(x, t) \pm \frac{1}{n}\frac{\partial \lambda^{\mp}(x)}{\partial x} + \frac{1}{2n^2}\frac{\partial^2 \lambda^{\mp}(x)}{\partial x^2}$$

Plugging these approximations into Eq. (9.38) and collecting terms proportional to $1/n$ gives

$$\begin{aligned}\frac{1}{n}\Big(- \frac{\rho(x, t)}{\partial x}\lambda^{+}(x) - \rho(x, t)\frac{\partial \lambda^{+}(x)}{\partial x} \\ + \rho(x, t)\frac{\partial \lambda^{-}(x)}{\partial x} + \frac{\partial \rho(x, t)}{\partial x}\lambda^{-}(x)\Big) \\ = -\frac{1}{n}\frac{\partial}{\partial x}\rho(x, t)(\lambda^{+}(x) - \lambda^{-}(x))\end{aligned}$$

Similarly collecting terms for $1/n^2$ gives

$$\frac{1}{2n^2}\frac{\partial^2}{\partial x^2}\rho(x, t)(\lambda^{+}(x) + \lambda^{-}(x))$$

so that we finally arrive at

$$\begin{aligned}\frac{\partial \rho(x, t)}{\partial t} &= -\frac{\partial}{\partial x}\rho(x, t)(\lambda^{+}(x) - \lambda^{-}(x)) \\ &+ \frac{1}{2}\frac{\partial^2}{\partial x^2}\rho(x, t)\frac{(\lambda^{+}(x) + \lambda^{-}(x))}{n}\end{aligned} \tag{9.39}$$

which is the Fokker-Planck equation. It describes the dynamics of the probability distribution $\rho(x, t)$ throughout time. The first term is usually called the drift term (indicating directionality) and the second term the diffusion term (indicating a random diffusion in all directions). Now letting $n \to \infty$, we obtain

$$\frac{\partial \rho(x, t)}{\partial t} = -\frac{\partial}{\partial x}\rho(x, t)(\lambda^{+}(x) - \lambda^{-}(x)). \tag{9.40}$$

We use as initial condition $\rho(x, t) = \delta(x - x_0)$ the Dirac delta function, so there the initial relatively frequency of type A players is x_0 and there is no uncertainty with respect to the initial condition. Indeed then $\rho(x, t) = \delta(x - x(t))$ for all time t. By definition of the Dirac delta function we obtain $\int x\rho(x, t)dx = \int x\delta(x - x(t))dx = x(t)$ so that

$$\frac{\partial x(t)}{\partial t} = \int_0^1 x \frac{\partial \rho(x,t)}{\partial t} dx \tag{9.41}$$

$$= - \int_0^1 x \frac{\partial}{\partial x} \delta(x - x(t))(\lambda^+(x) - \lambda^-(x))dx \tag{9.42}$$

and by partial integration we obtain that

$$\dot{x} = \lambda^+(x) - \lambda^-(x). \tag{9.43}$$

Working out $\lambda^{\pm}(x)$ for the Moran process yields

$$\lambda^-(x) = x(1-x)\frac{f_B(x)}{xf_A(x) + (1-x)f_B(x)}$$
$$\lambda^+(x) = x(1-x)\frac{f_A(x)}{xf_A(x) + (1-x)f_B(x)}$$

with finesses indicated as in Eq. (9.13) this leads to

$$\dot{x} = \frac{x(1-x)(F_A(x) - F_B(x))}{\frac{1}{\beta} - 1 + xF_A(x) + (1-x)F_B(x)} \tag{9.44}$$

which yields an adjusted replicator equation with fitness equal to payoff [34]. For the pairwise comparison, working out yields

$$\lambda^+(x) - \lambda^-(x) = x(1-x)\frac{\exp \beta(F_A - F_B) - 1}{\exp \beta(F_A - F_B) + 1},$$

which is the definition of the tangent hyperbolic, resulting in

$$\dot{x} = x(1-x)\tanh\left(\frac{\beta}{2}(F_A - F_B)\right). \tag{9.45}$$

For small β this reduces to exactly the replicator equation when taking the first order Taylor expansion, with fitnesses equal to payoff [35]. The first order approximation of the tangent hyperbolic function yields $\tanh \beta/2\Delta = \beta/2\Delta$ so that we arrive at

$$\dot{x} = \frac{\beta}{2}x(1-x)(F_A - F_B),$$

which is exactly the replicator for a rescaled time $t' = t\beta/2$. Hence, the replicator equation is consistent with the finite population models. Moreover, for 2×2 games a strategy is risk dominant if $x^* < 1/2$, while is has an evolutionary advantage ($\rho_A > 1/n$) if $x^* < 1/3$ [22].

9.1.4 Prisoner's Dilemma

Let us briefly look at the prisoner's dilemma, which corresponds to a payoff matrix with

$$A = \begin{array}{c} \\ C \\ D \end{array} \!\! \begin{array}{c} \begin{array}{cc} C & D \end{array} \\ \begin{pmatrix} R & S \\ T & P \end{pmatrix} \end{array}. \tag{9.46}$$

The prisoner's dilemma corresponds to $T > R > P > S$. Here R (reward) corresponds to the payoff when both agents cooperate, while if the one agents defects and the other cooperates he receives T (temptation) and the other receives S (sucker), while if both defect, both get a payoff of P (punishment). For our earlier example we had $R = -1$, $T = 0$, $S = -5$ and $P = -4$.

Let us start by looking at the Nash equilibrium and the ESS. Clearly, the strategy $x = (0, 1)$ (always defect) is a (strict) Nash equilibrium, which can be easily seen from

$$x^\top Ax = (0, 1) \begin{pmatrix} R & S \\ T & P \end{pmatrix} \begin{pmatrix} 0 \\ 1 \end{pmatrix} = P,$$

while for $y = (p, 1 - p) \neq x$ we obtain

$$y^\top Ax = (p, 1 - p) \begin{pmatrix} R & S \\ T & P \end{pmatrix} \begin{pmatrix} 0 \\ 1 \end{pmatrix} = Sp + (1 - p)P,$$

and since $S < P$, we obtain that $y^\top Ax < x^\top Ax$, and hence x is a strict Nash equilibrium, hence it is also an ESS. In fact, it is also the unique Nash equilibrium (and ESS). Let $y = (p, 1 - p)$ be any other strategy. Then

$$y^\top Ay - x^\top Ax = p^2(R - T) + p(1 - p)(S - P) < 0$$

because $T > R$ and $P > S$.

Let us look at this from the viewpoint of a finite population. Assume we have n agents, of two types only: those who always cooperate (AllC) and those who always defect (AllD). Let us assume there are n_C cooperators and n_D defectors. The payoff for a cooperator is then

$$F_C(\vec{n}) = R\frac{n_C - 1}{n - 1} + S\frac{n_D}{n - 1}$$

since if a cooperator is playing against a cooperator the payoff is R and against a defector it is S. For a defector the payoff is then

$$F_D(\vec{n}) = T\frac{n_C}{n - 1} + P\frac{n_D - 1}{n - 1}.$$

The difference in payoff is then

$$F_C(\vec{n}) - F_D(\vec{n}) = \frac{n_C}{n-1}(R-T) + \frac{n_D}{n-1}(S-P) + \frac{1}{n-1}(P-R),$$

which is always negative because $T > R > P > S$, and so $F_C(\vec{n}) < F_D(\vec{n})$, consistent with the Nash equilibrium. We hence expect defectors to generally win. The average payoff is then

$$\langle F \rangle = F_D(\vec{n})\frac{n_d}{n} + F_C(\vec{n})\frac{n_c}{n}.$$

Let us analyse the fixation probability ρ_C of a single cooperator mutant in a population of defectors. Let us first focus on the ratio ρ_C/ρ_D to investigate which of the strategies is risk dominant. By Eq. (9.19) cooperators are risk dominant if

$$R(n-2) + Sn > P(n-2) + Tn,$$

which for large n is $R+S > P+T$. This is neither the case for finite population size nor for large n given that $T > R > P > S$. Cooperators have an evolutionary advantage whenever $\rho_A > 1/n$ which by Eq. (9.24) is the case for large n if $2S + R > 2P + T$, which contradicts $T > R > P > S$, so that cooperators indeed never have an evolutionary advantage.

The replicator equation is then

$$\begin{aligned}\dot{x} &= x(1-x)(Rx + S(1-x) - Tx - P(1-x)) \\ &= x(1-x)((1-x)(S-P) + x(R-T)),\end{aligned}$$

which might have a fixed point (besides $x^* = 0$ and $x^* = 1$) at

$$x^* = \frac{P-S}{R-T+P-S}.$$

Indeed, for this fixed point to exist we must have $0 < x^* < 1$, which is never the case. Hence, there are only two fixed points, only one of which is stable, namely $x^* = 0$. Hence, for any initial population, the population evolves towards only defectors.

An alternative specification of the prisoner's dilemma, which is also often employed is the following (Fig. 9.2). Each agent can cooperate by providing a benefit $b > 0$ to his partner at a cost of $c < b$ to himself. This corresponds to the following payoff matrix

$$A = \begin{matrix} & C & D \\ C & b-c & -c \\ D & b & 0 \end{matrix}. \tag{9.47}$$

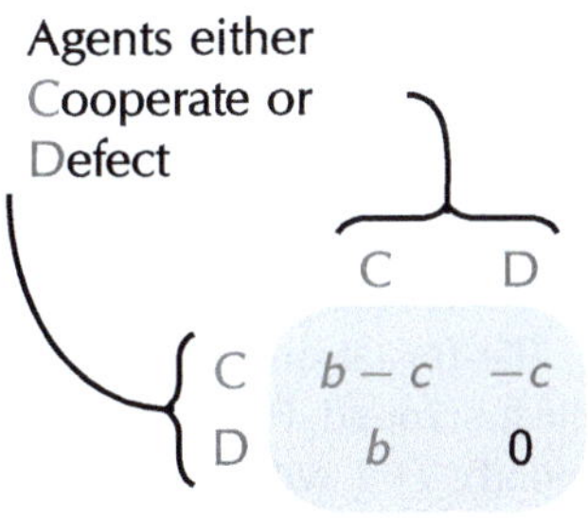

Fig. 9.2 Prisoner's dilemma

Again, only defecting is a strict Nash equilibrium and the only evolutionary stable strategy. Cooperating is risk dominant if

$$b - 2c > b \tag{9.48}$$

which is clearly never the case. The replicator equation becomes

$$\dot{x} = x(1-x)((b-c)x - c(1-x) - bx) = -cx(1-x) \tag{9.49}$$

so that only only $x^* = 1$ is a stable fixed point. Again, any initial population will evolve towards only defectors.

Hence, on all accounts cooperation is never evolutionary stable or advantageous. It is quite unlikely cooperators ever successfully invade a population of defectors and even less likely to become fixated in a population of defectors. On the other hand, defectors will easily invade a population of cooperators, and are likely to become fixated in the population. In the limit of large population size the evolutionary trend is always towards the defectors. In reality however, we do observe cooperation. Hence, the question is, how can cooperation ever have evolved?

9.2 Towards Cooperation

Various mechanisms for explaining the evolution of cooperation have been suggested [21], such as kin selection [9, 15], reciprocity [3] or group selection [38]. We will first explain briefly the concept of direct reciprocity and then introduce the concept of indirect reciprocity.

9.2.1 Direct Reciprocity

Direct reciprocity—or simply reciprocity for short—is based on the idea of "you scratch my back, I scratch yours". That is, if I help you, you should return the favour by helping me. This comes down to playing multiple rounds of the prisoner's dilemma. This is also known as the iterated prisoner's dilemma (IPD). There is a plethora of different strategies to play an IPD, but the one strategy that performs outstanding versus a wide array of different strategies is tit-for-tat (TFT), which embodies the principle of direct reciprocity [3].

Tit-for-tat simply does exactly what its opponent does and starts out cooperating on the first round. For example, if TFT plays against AllD (a player who defects on every round), TFT will cooperate on the first round, but will subsequently defect, because AllD defects. In this context it is also commonly demanded that $R > (T+P)/2$ so that cooperation has a higher reward than alternating between defecting and cooperating. Notice that for the specification using benefits b and costs c in Eq. (9.47) this is always the case. Let us use this specification to look how TFT evolves against AllD.

First we have to decide how many rounds k of the prisoner's dilemma there will be played. This is not exactly trivial. We know that for a single round only defecting is stable. So, if we know how many rounds there will be, it should be better to defect on the last round. But then the second to last round can be considered as the last round, and working our way back in that way, it should be better to always defect again. Hence, it is usually supposed that there is a certain probability w to have another round. This is sometimes poetically called "the shadow of the future".

The number of rounds then follows a geometric probability distribution, with an expected number of rounds of $\langle k \rangle = 1/(1-w)$. If a TFT player meets another TFT player, they continue to cooperate, since they start out cooperatively and then do what the other ones does (i.e. cooperate), and so both receive an expected payoff of $\langle k \rangle (b-c)$. If a TFT player plays against AllD, the TFT player will receive once a payoff of $-c$ after which he will also always defect, while the AllD player will then receive once a payoff of b. Two AllD players will never receive any payoff. The expected payoffs for TFT versus AllD is then as follows

$$A = \begin{array}{c} \\ \text{TFT} \\ \text{AllD} \end{array} \begin{array}{c} \begin{array}{cc} \text{TFT} & \text{AllD} \end{array} \\ \begin{pmatrix} \langle k \rangle (b-c) & -c \\ b & 0 \end{pmatrix} \end{array}. \tag{9.50}$$

Analysing when TFT is a strict Nash equilibrium, we arrive at

$$\frac{b}{c} > \frac{1}{w} \tag{9.51}$$

Hence, if the probability to continue another round is large enough, or the other way around if the benefit-cost ratio is large enough, cooperation has a chance to evolve. However, defecting will also always stay a Nash equilibrium since $-c < 0$. TFT is risk dominant if $\frac{b}{c} > \frac{2-w}{w}$, for large n. If we look at the fixation probability of TFT this is larger then neutral if $\frac{b}{c} > \frac{3-2w}{w}$.

One particular problem for TFT however is that in the face of errors it might end up defecting [19]. In order to analyse this let us look at what combination of cooperation and defecting two TFT players will end up. Let us denote by $q =$

$(q_{CC}, q_{CD}, q_{DC}, q_{DD})$ the probabilities to cooperate. Then q_{CC} is the probability to cooperate after a round in which the agent himself cooperated and his opponent cooperated, q_{CD} the probability the cooperate after a round in which the agent himself defected while his opponent defected, etc.... TFT then corresponds to the strategy $q = (1, 0, 1, 0)$.

Now let us look how two players with strategies q and q' end up in what type of situations. Let us denote by M the matrix of transition probabilities for the four different states CC, CD, DC and DD, so that for example $M_{CD,DC}$ represent the probability to move from state CD to state DC. This transition matrix is then

$$M = \begin{pmatrix} q_{CC}q'_{CC} & q_{CC}(1-q'_{CC}) & (1-q_{CC})q'_{CC} & (1-q_{CC})(1-q'_{CC}) \\ q_{CD}q'_{DC} & q_{CD}(1-q'_{DC}) & (1-q_{CD})q'_{DC} & (1-q_{CD})(1-q'_{DC}) \\ q_{DC}q'_{CD} & q_{DC}(1-q'_{CD}) & (1-q_{DC})q'_{CD} & (1-q_{CD})(1-q'_{CD}) \\ q_{DD}q'_{DD} & q_{DD}(1-q'_{DD}) & (1-q_{DD})q'_{DD} & (1-q_{DD})(1-q'_{DD}) \end{pmatrix}. \tag{9.52}$$

Let us denote the probability to be in a certain state with π. The probabilities then change according to $\pi(t+1) = M\pi(t) = M^t\pi(0)$. We are interested in the stationary state $\pi = \lim_t \pi(t) = M\pi$. If the game will be iterated long enough we will be in the stationary state and the payoff for a certain strategy q is then $F_q = \sum_s \pi_s A_s$. So, for two TFT players this becomes

$$M = \begin{array}{c} \\ CC \\ CD \\ CD \\ DD \end{array} \begin{array}{c} \begin{array}{cccc} CC & CD & DC & DD \end{array} \\ \begin{pmatrix} 1 & 0 & 0 & 0 \\ 0 & 0 & 1 & 0 \\ 0 & 1 & 0 & 0 \\ 0 & 0 & 0 & 1 \end{pmatrix} \end{array}.$$

From this one can observe that if somehow one of the two TFT players defects, that the two TFT players alternatively defect. They alternate then between states CD and DC, so that each take turn in defecting. In fact, let us suppose that with some probability ε a TFT player defects while he didn't intend to, which corresponds to strategy $q = (1-\varepsilon, 0, 1-\varepsilon, 0)$. The transition matrix is then

$$M = \begin{array}{c} \\ CC \\ CD \\ CD \\ DD \end{array} \begin{array}{c} \begin{array}{cccc} CC & CD & DC & DD \end{array} \\ \begin{pmatrix} (1-\varepsilon)^2 & (1-\varepsilon)\varepsilon & \varepsilon(1-\varepsilon) & \varepsilon^2 \\ 0 & 0 & 1-\varepsilon & \varepsilon \\ 0 & 1-\varepsilon & 0 & \varepsilon \\ 0 & 0 & 0 & 1 \end{pmatrix} \end{array}.$$

This transition matrix M has a single absorbing state, so that for all $\varepsilon > 0$ two TFT players (with errors) will end up defecting at stationarity.

For this reason often also another strategy is considered, which is more stable in the presence of errors, namely the Win-Stay-Loose-Shift (WSLS) strategy [19]. The idea is that whenever the agent is doing well, it will continue to make its current choice, and if it is doing not so well it will switch. Without errors this corresponds to the strategy $q = (1, 0, 0, 1)$. This yields the transition matrix for two WSLS players

$$M = \begin{array}{c} CC \\ CD \\ CD \\ DD \end{array} \overset{\begin{array}{cccc} CC & CD & DC & DD \end{array}}{\begin{pmatrix} 1 & 0 & 0 & 0 \\ 0 & 0 & 0 & 1 \\ 0 & 0 & 0 & 1 \\ 1 & 0 & 0 & 0 \end{pmatrix}}.$$

This matrix has as the state CC as the single absorbing states and hence two WSLS players are expected to end up cooperating, regardless of the initial condition.

Let us see how the WSLS strategy does in the face of errors. This then corresponds to the strategy $q = (1 - \varepsilon, 0, 0, 1 - \varepsilon)$. The corresponding transition matrix is then

$$M = \begin{array}{c} CC \\ CD \\ CD \\ DD \end{array} \overset{\begin{array}{cccc} CC & CD & DC & DD \end{array}}{\begin{pmatrix} (1-\varepsilon)^2 & (1-\varepsilon)\varepsilon & \varepsilon(1-\varepsilon) & \varepsilon^2 \\ 0 & 0 & 0 & 1 \\ 0 & 0 & 0 & 1 \\ (1-\varepsilon)^2 & (1-\varepsilon)\varepsilon & \varepsilon(1-\varepsilon) & \varepsilon^2 \end{pmatrix}}.$$

The stationary probability to cooperate is then

$$\pi_{CC} = \frac{(1-\varepsilon)^2}{2\varepsilon(1-\varepsilon)+1} \approx 1 - 4\varepsilon$$

which stays near 1 for small ε so that indeed the WSLS strategy stays cooperative when faced with small errors. Hence, the WSLS strategy is then quite robust with respect to these type of errors [19].

In fact, for the iterated prisoner's dilemma with an infinite number of rounds there exists a strategy that dominates all other strategies, the so-called zero-determinant strategies [30]. However, these zero-determinant players do not necessarily always cooperate amongst each other, and are not evolutionary stable [11]. For example the WSLS strategy actually obtains a better payoff, because they end up cooperating with each other [1].

9.2.2 *Indirect Reciprocity*

Humans have a tendency however to also cooperate in contexts beyond kin, group or repeated interactions. It is believed that some form of indirect reciprocity can explain the breadth of human cooperation [25]. Whereas in direct reciprocity the favour is returned by the interaction partner, in indirect reciprocity the favour is returned by somebody else, which usually involves some reputation. It has been theorized that such a mechanism could even form the basis of morality [2]. Additionally, reputation (and the fear of losing reputation) seems to play in important role in maintaining social norms [6–8].

In indirect reciprocity often a slightly different game is played, namely that of donation [25]. Instead of that a pair interacts both ways, one agent is selected to be the donor and the other the recipient. The donor may decide to give a benefit b to the recipient at a cost of c to himself. Hence, the recipient cannot immediately return the favour. Assuming random pairing, it will take some while before the same pair is chosen again for interaction, so that this reduces the possibilities for direct reciprocity. Nonetheless, this game reduces to our earlier game. So, even though we may speak of donor and recipient from time to time, the underlying game remains the same.

Usually indirect reciprocity is modelled using some form of reputation, which is often assume to be binary: agents are either good (G) or bad (B). We will consider indirect reciprocity of increasing complexity. The first order scheme is only based on the action of the donor: cooperate or defect. The second order scheme takes into account the reputation of the recipient. For example, it might be good to defect against a bad agent whereas this would be bad against a good agent. The third order scheme also takes into account the reputation of the donor. After all, perhaps agents should only care about their own reputation, not about the reputation of others.

First Order

In the simplest framework, the first order scheme, we assume that agents cooperate with “good” agents and defect with “bad” agents [24], since the other way around does not make much sense. This is similar to the so-called image scoring strategy [23]. If an agent cooperates he gets a good reputation, and if an agent defect he gets a bad reputation. In this simplest model, we obtain the following dynamics.

Let $r_i(t) \in \{-1, 1\}$ be the reputation of agent i in round t, so that $r_i(t) = -1$ denotes a bad reputation and $r_i(t) = 1$ denotes a good reputation. Then we randomly select an agent i for donation to an agent j. The reputation of agent j is $r_j(t)$ and if $r_j(t) = 1$ agent i will help and his reputation will become $r_i(t + 1) = 1$, but if $r_j(t) = -1$ agent i will defect because j has a bad reputation, but he also gets a bad reputation because he has defected and so $r_i(t+1) = -1$. In short, $r_i(t+1) = r_j(t)$. Let us assume all players have initially a good reputation so that $r_i(0) = 1$ for all i. Then it is clear that $r_i(t) = 1$ for all t and all i in the absence of any other players.

This is summarized in Fig. 9.3. Here the upper table denotes what the new reputation of the donor will be given his action. In this case this does not depend on the reputation of the recipient. Regardless of whether the recipient has a good or bad reputation, if the donor defects he will get a bad reputation. The lower table denotes what action the donor should take given the reputation of the recipient.

Now let us introduce some defectors. Let us assume there is a proportion of $x_D = n_D/n$ defectors and $x_C = n_C/n$ discerning cooperators (those that cooperate or not based on the reputation). Since a defector will defect by default, he will always get a bad reputation if he is chosen as a donor. A discerning cooperator will then get a bad reputation if he defects against anybody, including defectors. Let us denote by $n_C^G(t)$ the number of good agents amongst cooperators and by $n_D^G(t)$ the

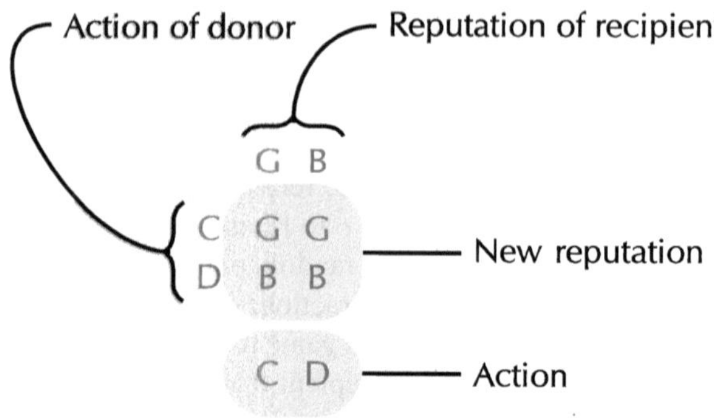

Fig. 9.3 First order indirect reciprocity

number of good agents amongst defectors. We denote by $p_C(t) = n_C^G(t)/n_C$ the probability a cooperator has a good reputation, and similarly so for the defectors denote $p_D(t) = n_D^G(t)/n_D$. The probability that the number of good agents amongst cooperators changes is then

$$\lambda^- = \Pr(\Delta n_C^G(t) = -1) = x_C p_C \left((1 - p_C)x_C + (1 - p_D)x_D\right)$$
$$\lambda^+ = \Pr(\Delta n_C^G(t) = 1) = x_C(1 - p_C)\left(p_C x_c + p_D x_D\right).$$

Then letting $n \to \infty$ [similar to the derivation of Eq. (9.39)] we obtain the reputation dynamics

$$\dot{p}_C = \lambda^+ - \lambda^- = (p_D - p_C)x_C x_D. \tag{9.53}$$

Defectors can only get a bad reputation, and never regain a good reputation since they will never cooperation. Working out similarly we obtain that

$$\dot{p}_D = -x_D p_D \tag{9.54}$$

The proportion of defectors that have a good reputation thus exponentially goes to zero, and $p_D(t) = e^{-x_D t}$. Substituting this solution in Eq. 9.53 we obtain as solution

$$p_C(t) = \frac{1}{1 - x_c} e^{-x_C x_D t} - \frac{x_c}{1 - x_c} e^{-x_D t}, \tag{9.55}$$

and so also $p_C(t) \to 0$ for $t \to \infty$. So, even though these cooperators discern quite well the defectors (the probability they have a good reputation quickly goes to zero), the cooperators themselves will also end up with a bad reputation. Hence, in the end there won't be any cooperation amongst these cooperators.

If we introduce unconditional cooperators (those who always cooperate irrespective of the reputation), the average reputation of the discerning cooperators will be higher. This still leads to problems however since the discerning cooperators and unconditional cooperators are evolutionary neutral, which might lead to invasion of defectors still [24].

Second Order

Given that the first order scheme is unable to maintain a high reputation for cooperators themselves, it makes sense to discern whether somebody defects against somebody of a bad reputation or not [4]. After all, if somebody defects against somebody with a bad reputation this should not be punished by giving that person also a bad reputation. This was also suggested in the literature and is similar to the standing strategy [14, 28], although experimental evidence suggested the simpler first order scheme (image scoring) prevailed among humans [17].

So, in the second order scheme we take into account the reputation of the recipient in order to determine whether cooperation or defection is justified. In particular, it allows to keep a good reputation when defecting against somebody with a bad reputation.

Denote again by $p_C = n_C^G/n_C$ the proportion of agents with a good reputation amongst the discriminating cooperation and $p_D = n_D^G/n_D$ the same among defectors. We denote by $K_{xy} \in \{0, 1\}$ if the new reputation should be good ($K_{xy} = 1$) or bad ($K_{xy} = 0$) given the reputation of the recipient x (good or bad) and the action y (cooperate or defect). For example, if $K_{GC} = 1 = G$ this indicates if an agent cooperates with an agent of a good reputation, he should get a good reputation. We denote by Z_x if one should cooperate or defect in a certain situation. For example, if $Z_G = 1 = C$ then agents should cooperate with those with a good reputation. We abbreviate $K_{xZ_x} = K_x$. The change rates are then given by

$$\begin{aligned}
\Pr(\Delta n_C^G = -1) &= x_C p_C \big((x_C p_C + x_D p_D)(1 - K_G) \\
&\quad + (x_C(1 - p_C) + x_D(1 - p_D))(1 - K_B)\big) \\
\Pr(\Delta n_C^G = 1) &= x_C(1 - p_C)\big((x_C p_C + x_D p_D)(1 - K_G) \\
&\quad + (x_C(1 - p_C) + x_D(1 - p_D))(1 - K_B)\big)
\end{aligned}$$

leading to

$$\dot{p}_C = (x_C p_C + x_D p_D) x_C (K_G - K_B) + x_C (K_B - p_C). \tag{9.56}$$

We would like to have that $p_C = 1$ is a fixed point of these dynamics, so that if we start with a good reputation for the cooperators that the cooperators retain a good reputation. This implies that

$$(x_C + x_D p_D)(K_G - K_B) + (K_B - 1) = 0.$$

If $K_B = 0$ we obtain that p_C is only a fixed point for specific values of x_C, x_D and p_D. Hence, in order for the fixed point to exist for all values, we must have $K_G = K_B = 1$. Obviously, we would also like discerning cooperators to cooperate, so that $Z_G = C$ and we obtain that $K_{GC} = G$. Suppose that $Z_B = C$, then an agent would also cooperate with somebody that has a bad reputation, hence, there is no

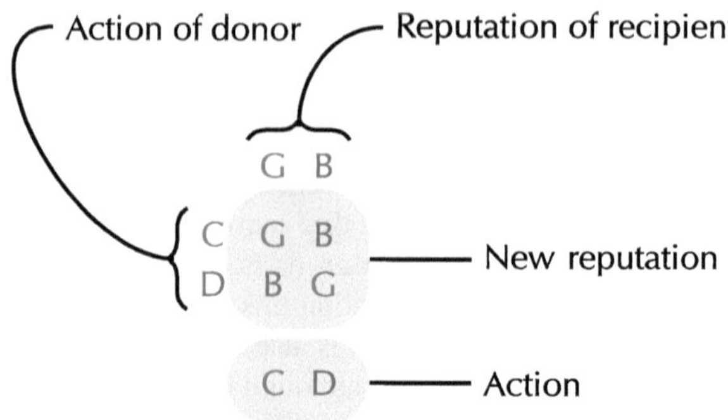

Fig. 9.4 Second order indirect reciprocity

interest in having a good reputation. So, we should have $Z_B = D$ and $K_{BD} = G = 1$. Furthermore, the cooperators should not cooperate with agents that have a bad reputation and so $K_{BC} = B = 0$. Likewise, if an agent defect against someone with a good reputation its reputation should be bad, so $K_{GD} = B = 0$ since otherwise defecting will not give you a bad reputation and others continue to cooperate despite the fact that you have a bad reputation.

Hence, among the $2^4 = 16$ possible strategies, only one seems viable against defectors. This strategy is denoted in Fig. 9.4. Again, the upper table represents the new reputation of the donor based on his action and the reputation of the recipient. In this case, it is only good to cooperate with a good agent and to defect with a bad agent.

This then gives us the reduced dynamics

$$\dot{p}_C = x_C(1 - p_C). \tag{9.57}$$

The solution of which is clearly independent of p_D, and leads to

$$p(t) = 1 - (1 - p(0))e^{-x_C t}. \tag{9.58}$$

Now let us look at whether this fixed point is stable. The derivative at that point is

$$\frac{\partial \dot{p}_C}{\partial p_C} = (x_C + x_D p_D)x_c(K_G - K_B) - x_c$$

which with $K_G = K_B = 1$ becomes $-x_C \leq 0$, and so the fixed point is stable.

Now let us look at the defectors. We denote by $K'_x = K_{xD}$. We obtain then

$$\dot{p}_D = (x_C p_C + x_D p_D)x_D(K'_G - K'_B) + x_D(K'_B - p_D). \tag{9.59}$$

Preferably we would like to have $p_D = 0$ a fixed point. This implies that

$$(x_C p_C + x_D)(K'_G - K'_B) + K'_B = 0. \tag{9.60}$$

We already know that $K_{BD} = G = 1$. Hence, this equality cannot hold for general x_C, p_C and x_D. So, we cannot expect all defectors to have a bad reputation.

Now let us see what is a feasible fixed point given $K'_B = 1$ and $K'_G = 0$. In that case the dynamics reduce to

$$\dot{p}_D = x_D(1 - p_D) - (x_C p_C + x_D p_D)x_D. \tag{9.61}$$

The only fixed point is thus

$$p_D^* = \frac{1 - x_C p_C}{1 + x_D}$$

which for $p_C = 1$ (the globally attracting fixed point for the cooperators) comes down to

$$p_D^* = \frac{x_D}{1 + x_D}. \tag{9.62}$$

The explicit solution for $p_D(t)$ is then

$$p_D(t) = p_D^* \left[1 - e^{-p_D^* t} - (1 - p(0))\, e^{-x_C t}\right] + p_D(0) e^{-p_D^* t}, \tag{9.63}$$

with p_D^* as in Eq. (9.62). Indeed $p_C(t) \to 1$ and $p_D(t) \to p_D^*$ for $t \to \infty$ for all initial conditions.

Assuming that $p_C(0) = p_D(0) = 1$, we obtain that $p_C(t) = 1$ for all t and that

$$p_D(t) = p_D^* \left(1 - e^{-p_D^* t}\right) + e^{-p_D^* t}.$$

Assuming finitesmal games the cumulative payoffs after time t are then as follows

$$F_C(x_C, t) = \frac{1}{t} \int_0^t b x_C p_C(\tau) - c(x_C p_C(\tau) + x_D p_D(\tau)) d\tau$$

$$F_D(x_C, t) = \frac{1}{t} \int_0^t b x_D p_D(\tau) d\tau.$$

Working this out we obtain

$$F_C(x_C, t) = x_c(b - c) - c x_D p_D^* - \frac{c}{t}\left(1 - e^{-p_D^* t}\right)$$

$$F_D(x_C, t) = b x_D p_D^* + \frac{b}{t}\left(1 - e^{-p_D^* t}\right).$$

For $t \to \infty$ then of course cooperators cooperate with defectors about p_D^* of the time, and we then obtain

$$F_C(x_C) = x_c(b-c) - cx_D p_D^*$$
$$F_D(x_D) = bx_D p_D^*$$

The difference can be simplified to

$$F_D(x_D) - F_C(x_D) = \frac{1}{1+x_D}\left(2bx_D^2 - (b-c)\right)$$

and the replicator equation becomes

$$\dot{x}_D = x_D(1-x_D)\frac{1}{1+x_D}\left(2bx_D^2 - (b-c)\right).$$

The fixed point in the replicator equation between the discerning cooperators and unconditional defectors is then

$$x_D^* = \sqrt{\frac{1}{2}\left(1-\frac{c}{b}\right)}. \tag{9.64}$$

Let us look at the stability of the fixed points. The derivative is

$$\begin{aligned}\frac{\partial \dot{x}_D}{\partial x_D} &= (1-2x_D)(F_D(x_D) - F_C(x_D)) \\ &+ \frac{x_D}{(1+x_D)^2}(2bx_D(2-x_D) + (b-c)x_c)\end{aligned} \tag{9.65}$$

For $x_D^* = 0$ (so $x_C^* = 1$) we obtain $\frac{\partial \dot{x}_D}{\partial x_D} = -(b-c) < 0$ equally for $x_D^* = 1$ (so $x_C^* = 0$) we obtain $\frac{\partial \dot{x}_C}{\partial x_D} = -b < 0$. Both fixed points are hence stable. For the fixed point x_D^* in Eq. (9.64) we obtain that only the latter term in derivative Eq. (9.65) is non-zero, and it is always positive. Hence, the point x_D^* is unstable. Since $c < b$ this fixed point $x_D^* > 0$ so that discriminating cooperators are never dominating. However, if $x_D^* > 2/3$ the discriminating cooperates are evolutionary advantageous, which is the case if $\frac{b}{c} > 9$, and they are risk dominant if $x_D^* > 1/2$ which gives $\frac{b}{c} > 2$.

Now let us look at the situation for relatively small t. We linearise around $t = 0$ and we then obtain

$$F_C(x_C, t) = x_C(b-c) - cp_D^*\left(1 + x_D - \frac{p_D^* t}{2}\right)$$
$$F_D(x_C, t) = bp_D^*\left(1 + x_D - \frac{p_D^* t}{2}\right).$$

In order to find the inner fixed point we have to solve a cubic polynomial, which isn't very informative. Instead, let us focus on when $F_C > F_D$ for $x_C = x_D = 1/2$

in order to determine when the cooperates are risk dominant. This is the case when

$$\frac{b}{c} > \frac{18}{t} - 1, \tag{9.66}$$

which is only valid for small t. Then similar to TFT the interaction should last long enough, or stated otherwise the benefit-cost ratio must be high enough. Furthermore, we can look at the evolutionary stability of the discerning cooperators. This means that at $x_D = 0$ the derivative should be $\frac{\partial \dot{x}_d}{\partial x_d} < 0$. This amounts to $\frac{\partial \dot{x}_d}{\partial x_d} = -(b-c) < 0$ similar as before. Hence, the discerning cooperators are still evolutionary stable for small times.

Third Order

One of the critiques against the second order scheme was that agents should only care about their own reputation and the associated rewards [14]. Hence, they shouldn't care about whether somebody else has a good reputation, but they should cooperate perhaps only to get a good reputation, and then defect as long as he keeps a good reputation. So, in the third order scheme we also take into account the reputation of the donor itself. This allows a plethora of different strategies, with 2^8 different new reputation schemes and 2^4 different possible action schemes [4, 26]. Nonetheless, only a few strategies make actual sense, similar to the second order scheme [27].

Let us then denote by $K(x, y, Z_{xy}) \in \{1, 0\}$ the reputation update function with $Z \in \{1, -1\}$ the action matrix, where x is the reputation of the donor and y the reputation of the recipient. The action matrix denotes whether an agent should cooperate or not in a certain situation. For example if $Z_{BG} = -1$ then an agent with a good reputation should defect with an agent with a bad reputation. So if $K(G, G, Z_{GG}) = G$ and $Z_{GG} = 1 = C$ this means if both i and j have a good reputation ($r_i = r_j = 1$) then i should cooperate ($Z_{GG} = 1$) and i keeps his good reputation $K(G, G, Z_{GG}) = 1 = G$. We will abbreviate $K(x, y, Z_{xy}) = K_{xy}$.

We are interested in knowing how different schemes (different reputation update matrices K and action matrices Z) will perform. As before, we will analyse the proportion of discerning cooperators that have a good reputation p_C and the proportion of defectors that have a good reputation p_D. We obtain

$$\begin{aligned}\Pr(\Delta n_C^G(t) = -1) = x_C p_C \big[&(x_C p_C + x_D p_D)(1 - K(G, G, Z_{GG})) \\ &+ (x_C(1-p_C) + x_D(1-p_D))(1 - K(G, B, Z_{GB}))\big]\end{aligned}$$

and

$$\begin{aligned}\Pr(\Delta n_C^G(t) = 1) = x_C(1-p_C) \big[&(x_C p_C + x_D p_D) K(B, G, Z_{BG}) \\ &+ (x_C(1-p_C) + x_D(1-p_D)) K(B, B, Z_{BB})\big]\end{aligned}$$

leading to

$$\dot{p}_C = (x_C p_C + x_D p_D)\,(x_C(1-p_C)K_{BG} - x_C p_C(1-K_{GG})) \\ + (x_C(1-p_C) + x_D(1-p_D))\,(x_C(1-p_C)K_{BB} - x_C p_C(1-K_{GB})) \,. \quad (9.67)$$

For defectors we obtain a similar result using $K'_{xy} = K(x, y, D)$

$$\dot{p}_D = (x_C p_C + x_D p_D)\left(x_D(1-p_D)K'_{BG} - x_D p_D(1-K'_{GG})\right) \\ + (x_C(1-p_C) + x_D(1-p_D))\left(x_D(1-p_D)K'_{BB} - x_D p_D(1-K'_{GB})\right) \,. \quad (9.68)$$

Notice that the only viable second order scheme corresponds to $Z_{\cdot B} = D$ and $Z_{\cdot G} = C$ and $K(\cdot,\cdot,C) = G$ and $K(\cdot,\cdot,D) = B$, which is indeed consistent with the results here.

Again, we would like that cooperators maintain a good reputation among each other. This means that $p{*}_C = 1$ should be a fixed point, and we arrive at

$$0 = \dot{p}_C = -x_C^2(1-K_{GG}) - x_C x_D(p_D(K_{GB} - K_{GG}) + 1 - K_{GB})$$

and by using $x_D = 1 - x_C$ and dividing by x_C we obtain

$$(K_{GG} - K_{GB})(x_c(1-p_D) + p_D) - (1 - K_{GB})) > 0$$

Notice that setting $x_c(1-p_D) + p_D = 0$ we obtain that

$$p_D^* = \frac{-x_C}{1 - x_C} < 0$$

which does not correspond to a valid solution. Hence, if $K_{GB} = 1$ we must have $K_{GG} = 1$. On the other hand, if $K_{GB} = 0$ then $K_{GG} = 1$ and $x_c(1-p_D) + p_D = 1$ so that $p_D = 0$. Hence, if there are no good defectors we can allow $K_{GB} = 0$. However, if there should be some perturbation so that there are some good defectors and $p_D > 0$ the fixed point shifts. So, in general we should set $K_{GB} = K_{GG} = 1$. Furthermore, $Z_{GG} = C$ because otherwise the population would have a high reputation, but would not cooperate. Also, $K(G, G, D) = K'_{GG} = 0 = B$ since otherwise defecting would not be "punished" by assigning a bad reputation. Furthermore, suppose that $Z_{GB} = C$. Then a good agent would cooperate with a bad agent, while not losing his good reputation. In that case it has no value to have a good reputation, and so $Z_{GB} = D$.

Again, demanding that the fixed point $p_C^* = 1$ is stable, we arrive at the following inequality for the derivative

$$0 \geq \left.\frac{\partial \dot{p}_C}{\partial p_C}\right|_{p_C=1} \\ = x_C^2(2K_{GG} - 2K_{GB} - K_{BG} + K_{BB}) - x_c(K_{BB} - K_{GB} + 1)$$

We already know that $K_{GG} = K_{GB} = 1$ so this reduces to $-K_{BG} - K_{BB}(1 - x_C) \leq 0$. If $K_{BG} = 0$ then $K_{BB} = 1$ or otherwise $\dot{p}_C^G = 0$ and the fixed point is not stable.

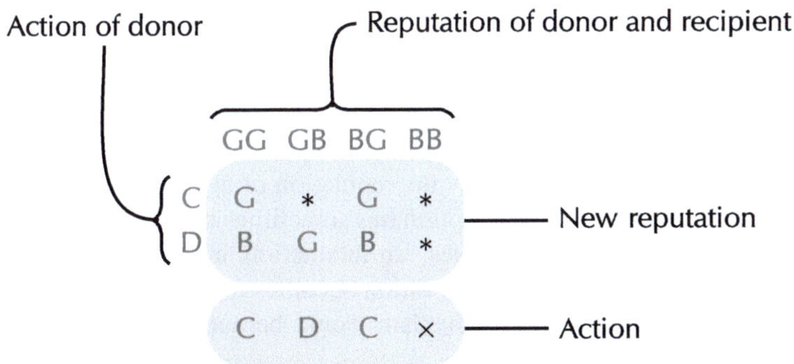

Fig. 9.5 Third order indirect reciprocity

However, we would like this stability to be especially the case for $x_C = 1$, and so we demand that at least $K_{BG} = 1$ in which case K_{BB} can be freely chosen. Now suppose that $Z_{BG} = -1 = D$. Then a bad player would get a good reputation by defecting against a good player, and would effectively reward defecting. Hence, $Z_{BG} = 1 = C$. Furthermore, suppose $K(B, G, D) = G$ then there would be no reason to cooperate to get a good reputation again, and so $K(B, G, D) = B$.

From this analysis we thus obtain the leading eight strategies indicated in Fig. 9.5. For a more elaborate argumentation as to why only these strategies perform well, refer to Ohtsuki and Iwasa [26, 27].

Besides demanding that the discerning cooperators have a good reputation, we could also demand that defectors will get a bad reputation. Hence, we demand that $p_D = 0$ is a fixed point, so that

$$x_C x_D p_C [K'_{BG} - K'_{BB}] + x_D K_{BB} = 0$$

If $K'_{BB} = 1$ then $K'_{BG} = 0$ and $x_C p_C = 1$, so that $x_C = 1$ and $p_C = 1$, and there are effectively no defectors. Hence, $K'_{BB} = 0$. If $K'_{BG} = 1$ then $x_C p_C = 0$ so that there are only defectors, in which case there are no relevant reputations. Hence $K'_{BB} = K'_{BG} = 0$, and this results in one of the leading eight strategies. If $K'_{BB} = 1$ we obtain that $p_D = \frac{x_D}{1+x_D}$ is a stable fixed point, assuming $p_C = 1$, consistent with the second order strategy. So, presumably $K'_{BB} = 0$ performs better in practice, because it allows to maintain a bad reputation for the defectors.

We can also analyse the stability of the fixed point $p_D = 0$. This point is stable whenever

$$x_D x_C p_C (K'_{GG} - K'_{GB}) - x_D(1 - K'_{GB}) \leq 0$$

From the analysis of the stability of the fixed point $p_C = 1$ we obtained that $K'_{GG} = 0$ and $K'_{GB} = 1$ so that indeed the fixed point 0 is stable.

9.3 Private Reputation

In the previous section we saw what type of strategies of indirect reciprocity can lead to cooperation. However, these reputations are usually considered as objective. That is, for an agent i all agents know the reputation of agent i, and all agents have the same view of agent i. This assumption has sometimes been relaxed by assuming only a part of the population "observes" an interaction, and updates their opinions accordingly. If this probability of observation becomes too small, reputations are no longer synchronized, and some mechanism would be necessary to maintain some coherence.

This is where we get back to dynamical models of social balance, which might overcome these issues. Although it allows to have private reputations (i.e. opinions), the dynamics also lead to some coherence. In addition, it models more explicitly the gossiping process, commonly suggested to be the foundation upon which reputations are forged. In addition, gossiping seems a more natural setting than "observing", and it was found to enhance cooperation in various experiments [29, 32, 33].

Notice that the model $\dot{X} = XX^{\top}$ is consistent with the leading second order strategy. Whenever the reputation of player j is good, (i.e. $X_{ij} > 0$) agents cooperate, and whenever it is bad (i.e. $X_{ij} \leq 0$) agents defect. Furthermore, it is considered good to defect against bad players. For example, suppose that j defects k ($X_{jk} \leq 0$) and that i thinks k is bad ($X_{ik} \leq 0$), then this will have a positive effect on j's reputation ($X_{ik}X_{jk} \geq 0$). Working out the remaining possibilities we arrive at the leading second order strategy illustrated in Fig. 9.4.

Although private reputations could also potentially follow a third order strategy, this seems unlikely. Already private reputations in a second order scheme are relatively demanding on people, and experimental evidence suggests that people commonly adopt simpler methods [17]. Hence, it makes sense to restrict ourselves to second order strategies. Furthermore, a private first order strategy is not that interesting for a private reputation, since everybody will always be judged in the same way, regardless of the reputation of the recipient.

Agents that use dynamics $\dot{X} = X^2$ will be referred to as type A, and agents that use dynamics $\dot{X} = XX^{\top}$ as type B. Defectors are agents that always defect. We assume that all agents talk, and share information truthfully as requested by the demanding party (including defectors). For example, if a type A agents asks a defector what he thinks of another agent of type B, he will answer that he has defected. If type B would have gossiped with a defector about an agent of type A, the defector would have replied what that agent of type A did.

In general, three types of agents might be in our population, and we can decompose the reputation matrix X accordingly as

$$X = \begin{pmatrix} X_A \\ X_B \\ X_D \end{pmatrix},$$

where X_A are the reputations in the eyes of agents of type A, X_B for type B and X_D for defectors. The reputation for defectors will not change, and will always be negative, i.e. $X_D(t) = X_D(0) < 0$. For the results displayed here we have used $X_D(0) = -10$, but results remain by and large the same when varying this parameter, as long as it remains sufficiently negative. As stated, the dynamics for type A and type B remain exactly as before

$$\dot{X}_A = X_A X,$$
$$\dot{X}_B = X_B X^\top,$$
$$\dot{X}_D = 0.$$

Agents defect whenever $X(t)_{ij} \leq 0$ and cooperate whenever $X(t)_{ij} > 0$. We define the cooperation matrix $C(t)$ accordingly

$$C_{ij}(t) = \begin{cases} 0 & \text{if } X_{ij} \leq 0 \\ 1 & \text{if } X_{ij} > 0 \end{cases}$$

Whenever an agent i cooperates with j, or $C(t)_{ij} = 1$, agent j receives a payoff of b at a cost of c to agent i, as illustrated in Fig. 9.2. Since we are dealing with continuous dynamics here, we assume the agents are involved in infinitesimally short games. Assuming the solution of $X_A(t)$ and $X_B(t)$ blows up at time t^*, we obtain the payoff vector P as

$$F = \frac{1}{n} \int_0^{t^*} bC(t)^\top e - cC(t)edt,$$

where $e = (1, \ldots, 1)$ the vector of all ones. Each element F_i contains the payoff for an individual agent i.

Based on the payoffs P we let the population evolve. We sample our new population based on the payoff of this old population. We define the replication probability as

$$\phi_i = \frac{\exp \beta F_i}{\sum_i \exp \beta F_i},$$

which is the Boltzmann probability distribution, where β represents the selective pressure. Higher β signifies higher selective pressure, and leads to a higher reproduction of those with a high payoff, and in the limit $\beta \to \infty$ only those with the maximum payoff reproduce. For $\beta \to 0$ this tends to the uniform distribution, where payoffs no longer play any role. In other words, we simulate a Wright-Fisher process (see Sect. 9.1.1) with fitness function $f_i = \exp \beta F_i$. We have used $\beta = 1$ as the "standard" selective pressure, but have also simulated for high selective pressure ($\beta = 5$) and low selective pressure ($\beta = 0.5$). We stop the simulation whenever one of the types becomes fixated in the population. We repeat this process 1,000 times for the results using $\beta = 1$, and for the low ($\beta = 0.5$) and high ($\beta = 5$) selective pressure 100 times in order to estimate the fixation probability. Finally, we initialize the population with an equal number of agents of each type. The initial reputation $X(0)$ is sampled from a standard Gaussian distribution with mean zero and standard deviation one. We reinitialize the reputation matrix every generation.

The results are displayed in Fig. 9.6 using a normalized cost of $c = 1$ (the ratio b/c drives the evolutionary dynamics). When directly competing against each other, type B has an evolutionary advantage (its fixation probability $\rho_B > 1/2$) compared to type A, already for relatively small benefits. When each type is playing against defectors (agents that always defect), type A seems unable to defeat defectors ($\rho_A < 1/2$) for any $b < 20$, while type B performs quite well against them.

When varying the number of agents, the critical benefit b^* at which type B starts to have an evolutionary advantage changes (i.e. where the fixation probability $\rho_B = 1/2$). For $b > b^*$ agents using the model $\dot{X} = XX^\top$ have a higher chance to become fixated, while for $b < b^*$ defectors tend to win. The inequality for type B to have an evolutionary advantage can be relatively accurately approximated by $b > b^* = \kappa\sqrt{n}$ where κ is estimated to be around $\kappa \approx 1.72 \pm 0.037$ (95 % confidence interval), see Fig. 9.7.

In summary, if $\frac{b}{c} > \kappa\sqrt{n}$ the model $\dot{X} = XX^\top$ has an evolutionary advantage. Type B is able to lead to cooperation and defeats type A. Based on these results, if a gossiping process evolved during the course of human history in order to maintain cooperation, the model $\dot{X} = XX^\top$ seems more likely to have evolved than $\dot{X} = X^2$. For smaller groups a smaller benefit is needed for the model $\dot{X} = XX^\top$ to become fixated. This dependence seems to scale only as $\sqrt{n}$, so that larger groups only need a marginally larger benefit in order to develop cooperation.

In addition to the results of competing both type A and type B separately against defectors, we also obtained results for populations initialized with type A, type B *and* defectors, all three at the same time. These results are largely the same as for one of the types against defectors. A small difference is that type A obtains a small advantage, because it can benefit from type B defeating the defectors. These results are reported in Fig. 9.8.

The results for the different selective pressure are reported in Fig. 9.9. A higher selective pressure leads to a higher evolutionary advantage for type B, as could be expected. A lower selective pressure levels the playing field, and allows type A to survive almost as frequently as type B, although still somewhat less frequently. The performance against defectors however remains largely unchanged for type A, and they are still unable to survive against defectors. For type B, they tend to win more frequently for low benefits b for low selective pressure, while for higher benefit b the high selective pressure allows them to thrive. This is probably due to the relatively slim evolutionary advantage of defectors versus type B for low b, while the advantage of type B players is quite substantial at high b.

In conclusion, for second order indirect reciprocity, there seems to be only one leading strategy, namely it is good to cooperate with good people and good to defect against bad people, as reported in Fig. 9.4. When considering private reputations, second order strategies make more sense than first or third order strategies. Considering the single leading second order strategy for private reputation gives rise to a

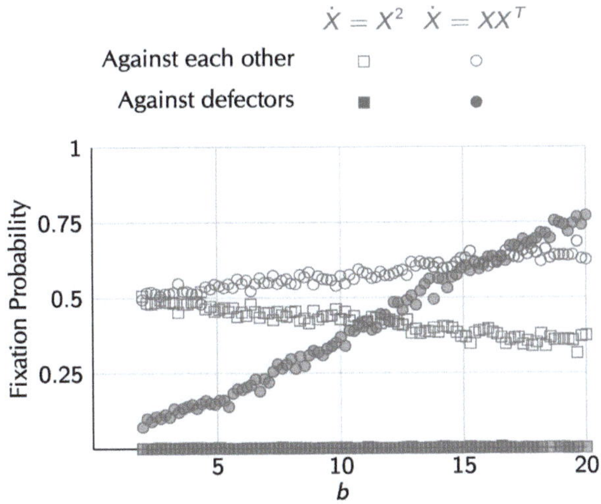

Fig. 9.6 Evolutionary performance of both models

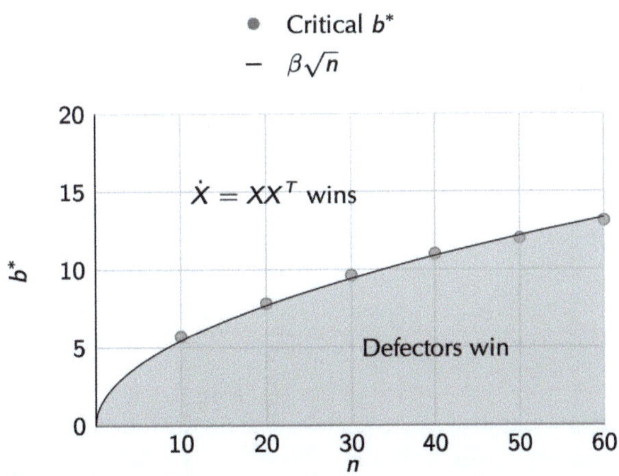

Fig. 9.7 Performance for different number of agents

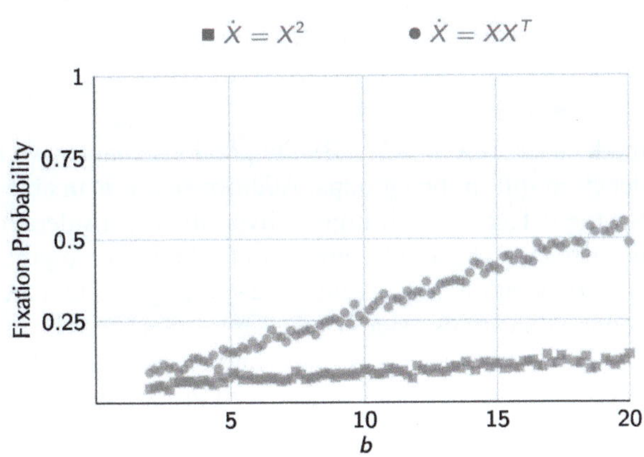

Fig. 9.8 Results including type A, B and defectors

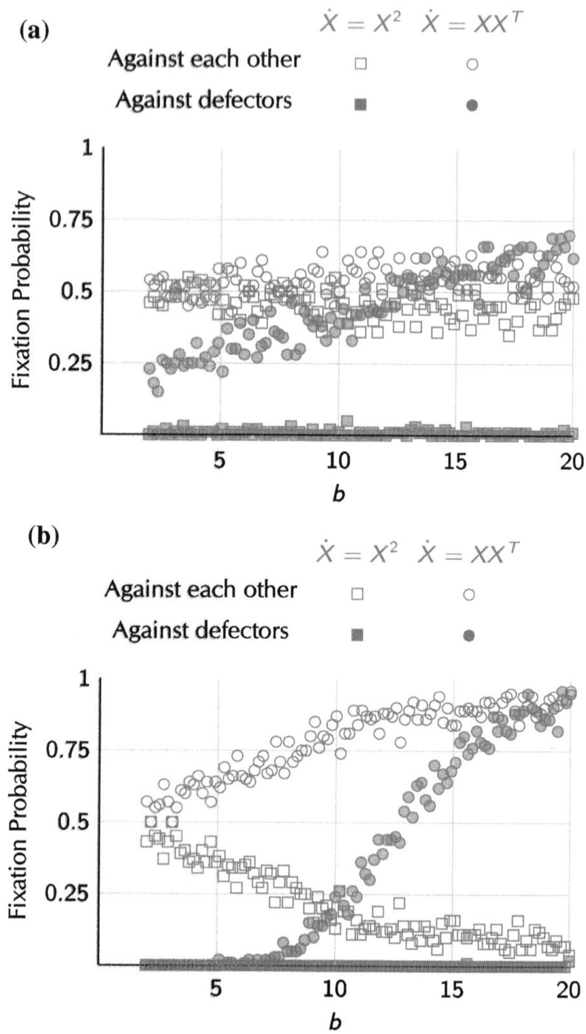

Fig. 9.9 Results different intensities of selection

model such as $\dot{X} = XX^{\top}$. By the previous chapter we know this model has the tendency to split in two groups. Additionally, it is unable to maintain a bad reputation for the defectors at all times. Given these considerations, it is possible that social balance emerges as a by-product of a indirect reciprocity scheme in order to maintain cooperation. In this sense, the splitting into two factions and the maintenance of cooperation are two sides of the same coin. It has been suggested that gossiping has evolved to maintain cooperation and cohesive social networks [5]. If that is true, its likely consequence is also that social groups split into antagonistic factions.

References

1. Adami C, Hintze A (2013) Evolutionary instability of zero-determinant strategies demonstrates that winning is not everything. Nat Commun 4. doi:10.1038/ncomms3193
2. Alexander RD (1987) The biology moral systems. Aldine de Gruyter, New York
3. Axelrod R (1984) Evolution of cooperation. Basic Books. ISBN 0465021220
4. Brandt H, Sigmund K (2004) The logic of reprobation: assessment and action rules for indirect reciprocation. J Theor Biol 231(4):475–486. doi:10.1016/j.jtbi.2004.06.032
5. Dunbar RIM (1998) Grooming, gossip, and the evolution of language. Harvard University Press, Cambridge ISBN 0674363361
6. Elias N, Scotson JL (1994) The established and the outsiders. SAGE Publications, London ISBN 9780803979499
7. Fehr E, Fischbacher U (2004) Third-party punishment and social norms. Evol Hum Behav 25(2):63–87. doi:10.1016/S1090-5138(04)00005-4
8. Friedkin NE (2001) Norm formation in social influence networks. Soc Netw 23(3):167–189. doi:10.1016/S0378-8733(01)00036-3
9. Hamilton W (1964) The genetical evolution of social behaviour I. J Theor Biol 7(1):1–16. doi:10.1016/0022-5193(64)90038-4
10. Hamilton WD (1967) Extraordinary sex ratios. Science 156(3774):477–488. doi:10.1126/science.156.3774.477
11. Hilbe C, Nowak MA, Sigmund K (2013) Evolution of extortion in iterated prisoner's dilemma games. Proc Nat Acad Sci 110(17):6913–6918. doi:10.1073/pnas.1214834110
12. Hofbauer J, Sigmund K (1998) Evolutionary games and population dynamics. Cambridge University Press, Cambridge. ISBN 052162570X
13. Imhof LA, Nowak MA (2006) Evolutionary game dynamics in a Wright-Fisher process. J Math Biol 52(5):667–681. doi:10.1007/s00285-005-0369-8
14. Leimar O, Hammerstein P (2001) Evolution of cooperation through indirect reciprocity. Proc R Soc Biol Sci 268(1468):745–753. doi:10.1098/rspb.2000.1573
15. Maynard Smith J (1982) Evolution and the theory of games. Cambridge University Press, Cambridge 0521288843
16. Maynard Smith J, Price G (1973) The logic of animal conflict. Nature 246(5427):15–18. doi:10.1038/246015a0
17. Milinski M, Semmann D, Bakker TC, Krambeck HJ (2001) Cooperation through indirect reciprocity: image scoring or standing strategy? Proc R Soc Biol Sci 268(1484):2495–2501. doi:10.1098/rspb.2001.1809
18. Nash J (1953) Two-person cooperative games. Econometrica 21(1):128. doi:10.2307/1906951
19. Nowak M, Sigmund K (1993) A strategy of win-stay, lose-shift that outperforms tit-for-tat in the Prisoner's Dilemma game. Nature 364(6432):56–58. doi:10.1038/364056a0
20. Nowak MA (2006a) Evolutionary dynamics: exploring the equations of life. Belknap Press of Harvard University Press. ISBN 0674023382
21. Nowak MA (2006b) Five rules for the evolution of cooperation. Science 314(5805):1560–1563. doi:10.1126/science.1133755
22. Nowak MA, Sasaki A, Taylor C, Fudenberg D (2004) Emergence of cooperation and evolutionary stability in finite populations. Nature 428. doi:10.1038/nature02360
23. Nowak MA, Sigmund K (1998a) Evolution of indirect reciprocity by image scoring. Nature 393(6685):573–577. doi:10.1038/31225
24. Nowak MA, Sigmund K (1998) The dynamics of indirect reciprocity. J Theor Biol 194(4):561–574. doi:10.1006/jtbi.1998.0775
25. Nowak MA, Sigmund K (2005) Evolution of indirect reciprocity. Nature 437(7063):1291–1298. doi:10.1038/nature04131
26. Ohtsuki H, Iwasa Y (2004) How should we define goodness?—reputation dynamics in indirect reciprocity. J Theor Biol 231(1):107–120. doi:10.1016/j.jtbi.2004.06.005
27. Ohtsuki H, Iwasa Y (2006) The leading eight: social norms that can maintain cooperation by indirect reciprocity. J Theor Biol 239(4):435–444. doi:10.1016/j.jtbi.2005.08.008

28. Panchanathan K, Boyd R (2003) A tale of two defectors: the importance of standing for evolution of indirect reciprocity. J Theor Biol 224(1):115–126. doi:10.1016/S0022-5193(03)00154-1
29. Piazza J, Bering JM (2008) Concerns about reputation via gossip promote generous allocations in an economic game. Evol Human Behav 29(3):172–178. doi:10.1016/j.evolhumbehav.2007.12.002
30. Press WH, Dyson FJ (2012) Iterated Prisoner's Dilemma contains strategies that dominate any evolutionary opponent. Proc Natl Acad Sci USA 109(26):10409–10413. doi:10.1073/pnas.1206569109
31. Sella G, Hirsh AE (2005) The application of statistical physics to evolutionary biology. Proc Natl Acad Sci USA 102(27):9541–9546
32. Sommerfeld RD, Krambeck HJ, Milinski M (2008) Multiple gossip statements and their effect on reputation and trustworthiness. Proc R Soc Biol Sci 275(1650):2529–2536. doi:10.1098/rspb.2008.0762
33. Sommerfeld RD, Krambeck HJ, Semmann D, Milinski M (2007) Gossip as an alternative for direct observation in games of indirect reciprocity. Proc Natl Acad Sci USA 104(44):17435–17440. doi:10.1073/pnas.0704598104
34. Traulsen A, Claussen J, Hauert C (2005) Coevolutionary dynamics: from finite to infinite populations. Phys Rev Lett 95(23):238701. doi:10.1103/PhysRevLett.95.238701
35. Traulsen A, Nowak M, Pacheco J (2006) Stochastic dynamics of invasion and fixation. Phys Rev E 74(1):011909. doi:10.1103/PhysRevE.74.011909
36. Traulsen A, Shoresh N, Nowak MA (2008) Analytical results for individual and group selection of any intensity. Bull Math Biol 70(5):1410–1424. doi:10.1007/s11538-008-9305-6
37. Von Neumann J, Morgenstern O (2007) Theory of games and economic behavior. Princeton University Press. ISBN 0691130612
38. Wilson DS (1975) A theory of group selection. Proc Natl Acad Sci 72(1):143–146. doi:10.1073/pnas.72.1.143

Chapter 10
Ranking Nodes Using Reputation

In the previous chapter we investigated the evolution of cooperation. In many situations we would like to know whether we could trust someone to cooperate or not. That is, suppose we are given some network of who (dis)trusts whom, would it then be possible to state whom we should trust and whom we should not trust? After all, perhaps somebody who indicates that he trusts somebody might not be trustworthy himself.

10.1 Ranking Nodes

Suppose for the moment we would have only indications of trust (i.e. only positive links). In what way could we then indicate which node should be trusted more so then others? This idea is known as ranking nodes according to some reputation (or trust). For example, this idea forms the core of Google's PageRank. It is the score Google assigns to pages indicating whether such a page has a "good" reputation or not, in order to return relevant search results [6].

The ranking of nodes, or assigning some "importance" or "trust" scores to nodes. Already in the 1970s, various researchers from the social sciences have introduced concepts such as betweenness [11], closeness [12] and eigenvector centrality [4, 5] to measure how central or important a node in the network was. For example, centrality-like measures are shown to play an important role in spreading processes on networks, such as failing cascades [25], or the infection process of sexually transmitted diseases [7, 8]. Furthermore, it helps to identify different roles nodes might play in a network [21].

In the 1990s several alternative ranking measures were added, notably Kleinbergs HITS-algorithm [15], and Google's PageRank [6]. When file sharing and especially peer-to-peer applications grew, these measures, and variants thereof, became popular to keep "good" peers in the sharing network, and exclude "bad" peers [1, 14]. Reputation and trust also plays a vital role in online markets such as eBay [22].

V. Traag, *Algorithms and Dynamical Models for Communities and Reputation in Social Networks*, Springer Theses, DOI: 10.1007/978-3-319-06391-1_10,

Negative links however, are usually not taken into account by these ranking measures, or worse, they break down when negative entries appear as weights of the links. However, the signs of links (positive or negative) should not be ignored, since they may bear important consequences for the structure of the network, not in the least for the ranking of nodes. Proposals have been made to include such semantic information in hyperlinks on the World Wide Web [18]. Negative links are also present in various other settings such as reputation networks [17], sharing networks [14], social networks [24] and international networks [16], and play a key, if not vital, role in these networks. Studying how negative links influence the importance of nodes may help the understanding of such systems, and such a concept of importance might facilitate the analyses of such networks.

Let us first briefly review the PageRank. Assume we have some adjacency matrix A. The reputation r_i of node i should then be higher when it is recommended by nodes of a higher reputation. On the other hand, if a page points to many pages, the endorsement should be less strong. In other words, we could define the reputation recursively as

$$r_i = \sum_j \frac{A_{ji}}{k_j} r_j, \tag{10.1}$$

and the reputation r_i of node i is the sum of the reputations of the nodes that point to node i proportional to the degree. If we set $M = (D^{-1}A)^\top$ with $D = \operatorname{diag}(k)$ with k the degrees, this can also be written as

$$r = Mr. \tag{10.2}$$

Notice that this is equivalent to the Markov chain for a random walk (see Sect. 2.2.6). Hence, the reputation has a nice interpretation: it is the probability that we visit that node during a random walk . Moreover, in the context of surfing on the web this is also pertinent. The reputation of a web page can then be regarded as the probability that a random surfer visits this page.

Although this forms the basis for ranking, it has two problems: (1) dangling nodes, i.e. nodes that have no outgoing links; and (2) unconnected graphs. The first problem is remedied by supposing that in the random walk, whenever one meets a site that has no outgoing links, we will chose another site at random. In order to do so let $a_i = 1$ if $k_i = 0$ and $a_i = 0$ otherwise. Then we define $A' = A + ae^\top$ where $e = (1, \ldots, 1)$. We then set $M = (D^{-1}A')^\top$ where $D = \operatorname{diag}(k)$ with $k = A'e$ the degrees of A'. The second problem is remedied by adding a uniform probability to go to any site at all times. So, at each step, there is some probability α that the random surfer randomly jumps to another website, sometimes called the zap factor, which is commonly set to $\alpha = 0.85$. This corresponds to setting

$$G = \alpha M + (1 - \alpha)\frac{1}{n}ee^\top. \tag{10.3}$$

The page rank is then defined as the vector x for which $x = Gx$, i.e. the dominant eigenvector of the matrix G. This matrix is usually called the Google matrix. Because of the zapping factor α, the dominant eigenvector is unique and convergence is relatively quick.

However, this and other methods only work if all weights $A_{ij} \geq 0$ are non-negative. Hence, for negative links these type of methods break down, and other methods are needed. We will now analyse how we can introduce such a method.

Recently there has been more attention to negative links in ranking measures, for example PageTrust [9]. The difference between PageTrust and PageRank is that in the random walk in PageTrust nodes that are negatively pointed to during the random walk are blacklisted, and are visited less often, thereby lowering their PageTrust score. Another suggestion was to first calculate some ranking using only the positive links (e.g. using PageRank), and then apply one step of distrust, so that the negative links are taken into account [13, 18].

It was also suggested to introduce a propagation of distrust [13], implying that if i distrusts j, and j distrusts r, then i should trust r (the adage that the enemy of my enemy is my friend). The authors noted that this could lead to situations in which a node is its own enemy (if one considers a cycle of three negative links), reminiscent of social balance (see Chap. 7).

Given that a signed network might not be strictly balanced, an enemy of an enemy is not necessarily a friend. That is, if a node has a negative reputation, his links should not be distrusted, only trusted less. In other words, we should not assume a node with a negative reputation is not trustworthy (if he points negatively towards someone, we should not interpret it as positive, and vice versa), we should only trust his judgements less. This will actually follow from the derivation of the measure based on a discrete choice argument, which we will present in the following section. Most of the existing algorithms dealing with negative links do not apply distrust in such a recursive manner, thereby limiting their effect. Furthermore, none of the algorithms can actually deal with negative reputations, while this negativity can actually provide additional insight. For example, a negative reputation would signal that such a node should be blocked from the network.

Interestingly, a slightly different formulation, namely that r_i is some opinion or belief that node i holds, and it is updated according to the beliefs of its neighbours as

$$r_i(t+1) = \sum_j \frac{A_{ij}}{k_j} r_j(t),$$

has quite a different behaviour. Under the same conditions as for Eq. 10.2 (see also Sect. 2.2.6) these opinions converge to some consensus, such that all agents have the same opinion or belief, i.e. $r_i(t) = r_j(t) = r = \langle r(0) \rangle$ for $t \to \infty$ for all agents [10, 20]. These results are independent of whether the opinions or beliefs are positive or negative. If negative links are considered however results do change, and this is subject of recent research [2, 23], but we will not consider it further in this thesis.

10.2 Including Negative Links

Let as usual $G = (V, E)$ be a directed graph with $n = |V|$ nodes and $m = |E|$ edges. Each edge (i, j) has an associated weight $w_{ij} \in \mathbb{R}$ which can possibly be negative. By A we denote the $n \times n$ weighted adjacency matrix associated to the graph, such that $A_{ij} = w_{ij}$ if there is an (i, j) edge and zero otherwise. Furthermore, let r_i be some reputation of node i (we will make this explicit later on). We consider the links to indicate a certain trust: if node i points positively (negatively) to node j, this indicates that i trusts (distrusts) j. The goal is to infer some global trust values from the local trust links.

Suppose we are asked which node to trust, if we were to choose one. We assume that a higher reputation indicates some degree of trust, so we should preferably choose nodes which have a high reputation r_i. However, there might be some errors in choosing the one with the highest reputation. This is where the framework of discrete choice theory comes in.

The usual background for discrete choice theory is the following [3]. Suppose there are n different choices (in our case, nodes), which have a different associated utility u_i. We observe the utility o_i and have some error term ϵ_i such that

$$u_i = o_i + \epsilon_i. \tag{10.4}$$

We would like to choose the object with the maximum utility. However, since we only observe o_i, it is uncertain which item actually has the maximum real utility. So, the question becomes: what is the probability we will select a certain object? That is, what is the probability that $u_i \geq u_j$ for all $i \neq j$, or

$$\Pr(u_i = \max_j u_j), \tag{10.5}$$

depending on the observed utility o_i and the error term ϵ_i. In our case, we equate the observed utility o_i with some reputation r_i. We assume the real reputation is then $u_i = r_i + \epsilon_i$, where ϵ_i is the error made in observing the reputation.

The probability of choosing the node with the highest reputation depends on the distribution of the error term ϵ_i. Using the following assumption for the error term, we arrive at the well known multinomial logit model [3]. Suppose the ϵ_i are i.i.d. double exponentially distributed[1] according to

$$\Pr(\epsilon_i \leq x) = \exp - \left[\exp - \left(\frac{x}{\mu} + \gamma \right) \right], \tag{10.6}$$

where $\gamma \approx 0.5772$ is Euler's constant. The mean of Eq. (10.6) equals zero, and the variance equals $1/6\pi^2\mu^2$. With this error distribution it can be proven [3] that the probability node i has the highest real reputation becomes

[1] This distribution is also known as the Gumbel distribution.

$$p_i = \frac{\exp \frac{r_i}{\mu}}{\sum_j \exp \frac{r_j}{\mu}}. \tag{10.7}$$

This probability distribution is known as the Boltzmann distribution The probability a node i has the highest reputation, increases with higher reputation r_i, depending on the amount of noise characterized by μ, which we will term the "uncertainty". There are two extreme scenarios depending on μ. If $\mu \to \infty$ the variance goes to infinity, and the contribution of the observed reputation in $u_i = r_i + \epsilon_i$ becomes negligibly small. In that case, the probability a node has the highest real reputation becomes uniform, or $p_i = 1/n$. In the other extreme, $\mu \to 0$, there is essentially no error, and we will always be correct in choosing nodes with a maximum r_i. That is, if there is a set of nodes M with $r_i = \max_j r_j$ for $i \in M$, then $p_i = 1/|M|$ for $i \in M$, and zero otherwise.

The probabilities p shows how much we should trust nodes. Nodes with a higher reputation are more trustworthy than nodes with a lower reputation. The difference in trust becomes more pronounced with decreasing μ, up to the point where we only trust nodes with the highest reputation. We shall call these probabilities the trust probabilities.

The trust probabilities p depend on the reputation r_i, which we will define now. We will ask a certain node j to provide the reputation values of the other nodes. That is, we ask node j to be the judge of his peers. Since we consider A_{ji} to be the trust placed by node j in node i, we will assume that if node j is the judge, he would simply say that $r_i = A_{ji}$. The general idea is that the probability to be a judge depends on the reputation, which then influences that probability again.

The probability to be chosen as judge is simply p_i. Using those probabilities p_i, we select a judge at random, and let him give his opinion on the reputation of his peers. We thus allow trustworthy nodes a higher probability to judge their peers. The expected reputation can then be written as

$$r_i = \sum_j A_{ji} p_j,$$

or in matrix notation,

$$r = A^\top p,$$

where $A^\top$ is the transpose of A and p is a column probability vector (i.e. $\|p\|_1 = 1$ and $p_i \geq 0$). If we plug this formulation of the reputation into Eq. (10.7) we obtain a recursive formulation of trust probabilities

$$p(t+1) = \frac{\exp \frac{1}{\mu} A^\top p(t)}{\| \exp \frac{1}{\mu} A^\top p(t) \|_1}, \tag{10.8}$$

for some initial condition $p(0)$, with $\exp(\cdot)$ the element-wise exponential. Notice that if we add some constant c to A, then p will remain unchanged. We will prove next that this iteration actually converges to a unique fixed point p^*, i.e. independent of the initial conditions, for some range of values for μ. The final values of the trust probabilities can thus be defined as the limiting vector $p^* = \lim_{t\to\infty} p(t)$ or, equivalently, the fixed point p^* for which

$$p^* = \frac{\exp \frac{1}{\mu} A^\top p^*}{\| \exp \frac{1}{\mu} A^\top p^* \|_1}, \tag{10.9}$$

and the final reputation values as

$$r^* = A^\top p^*. \tag{10.10}$$

Notice that these reputation values are also a fixed point of the equation

$$r^* = A^\top \frac{\exp \frac{1}{\mu} r^*}{\| \exp \frac{1}{\mu} r^* \|_1} \tag{10.11}$$

and that the trust probabilities are related to the reputation values as

$$p^* = \frac{\exp \frac{1}{\mu} r^*}{\| \exp \frac{1}{\mu} r^* \|_1}. \tag{10.12}$$

In this sense, the trust probabilities and the reputation values can be seen as a dual formulation of each other.

Upon closer examination of Eq. (10.11), a certain node j might indeed get a negative reputation, but his judgements are taken less into account, they are not reversed. That is, as soon as a node has a negative reputation, we do not assume he is completely untrustworthy, and that his negative judgements should be taken positive, but only that he is less trustworthy. This means we indeed do not assume that the enemy of my enemy is my friend. A node could get a negative reputation for example if he is negatively pointed to by trustworthy nodes. This approach can be summarized in the idea that the reputation of a node depends on the reputation of the nodes pointing to him, or stated differently, a node is only as trustworthy as the nodes that trust him. Notice that this idea is similar to that of PageRank, namely that nodes are as important or trustworthy as the neighbours pointing to him [6].

Let us take a look at a small example to see what the effect is of negative links in a network as shown in Table 10.1. There is only one negative link, from a to d. The effect of the negative link becomes more penalizing when μ is decreased, as shown in Table 10.1b. That has also consequences for node e, who is only pointed to by d, who receives little trust, which then also leads to little trust for e. The PageRank for these nodes (for which we did not take into account the negative link, and used a zapping factor of 0.85) are provided as comparison, which assigns nodes d and e actually higher rankings.

Table 10.1 Example trust probabilities. **a** Example network. **b** Trust for various values of μ. **c** Cyclic behaviour $\mu = 0$

(a)

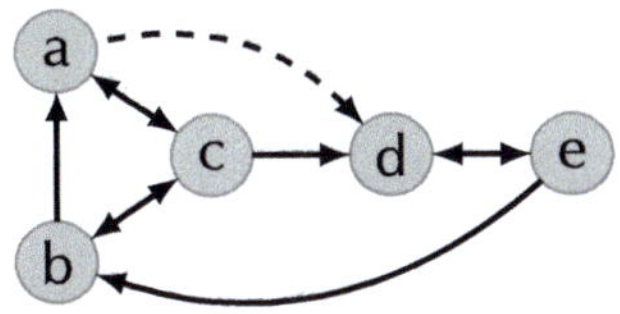

(b)

	PR	$\mu = 1$	$\mu = 1/5$	$\mu = 1/8$
a	0.183	0.223	0.384	0.424
b	0.184	0.213	0.179	0.142
c	0.263	0.223	0.384	0.424
d	0.184	0.171	0.026	0.005
e	0.186	0.171	0.026	0.005

(c)

	$t = 0$	$t = 1$	$t = 2$	$t = 3$
a	0.20	0.50	0.3	0.5
b	0.20	–	0.3	–
c	0.20	0.50	0.3	0.5
d	0.20	–	–	–
e	0.20	–	–	–

Of course, this measure can also be applied to networks without negative links. It is interesting to compare the exponential rank to the PageRank. In this case we have taken the co-authorship network of network scientists from [19]. This network includes 379 nodes in the largest connected component, and in Table 10.2 we list the top 10 highest ranked nodes for three different methods: (1) PageRank; (2) exponential rank with $\mu = 0.1$; and (3) exponential rank with $\mu = 1$. A famous network scientist, Barabási remains the highest ranked author in all three methods. For the rest there are quite some differences between PageRank and the exponential rank using $\mu = 0.1$. The rankings for $\mu = 0.1$ are relatively similar to the rankings for $\mu = 1$. Nonetheless, the correlation between the PageRank and the two different exponential rankings are quite high: 0.91 and 0.97 for $\mu = 0.1$ and $\mu = 1$ respestively. The rank correlation reveals there are more changes in the rank though, reaching only 0.61 for both $\mu = 0.1$ and $\mu = 1$. We visualize the network using PageRank in Fig. 10.1a and the exponential ranking with $\mu = 0.1$ in Fig. 10.1b.

We will now show that indeed this limit converges (for some range of μ) and is unique, i.e. does not depend on the actual initial condition $p(0)$.

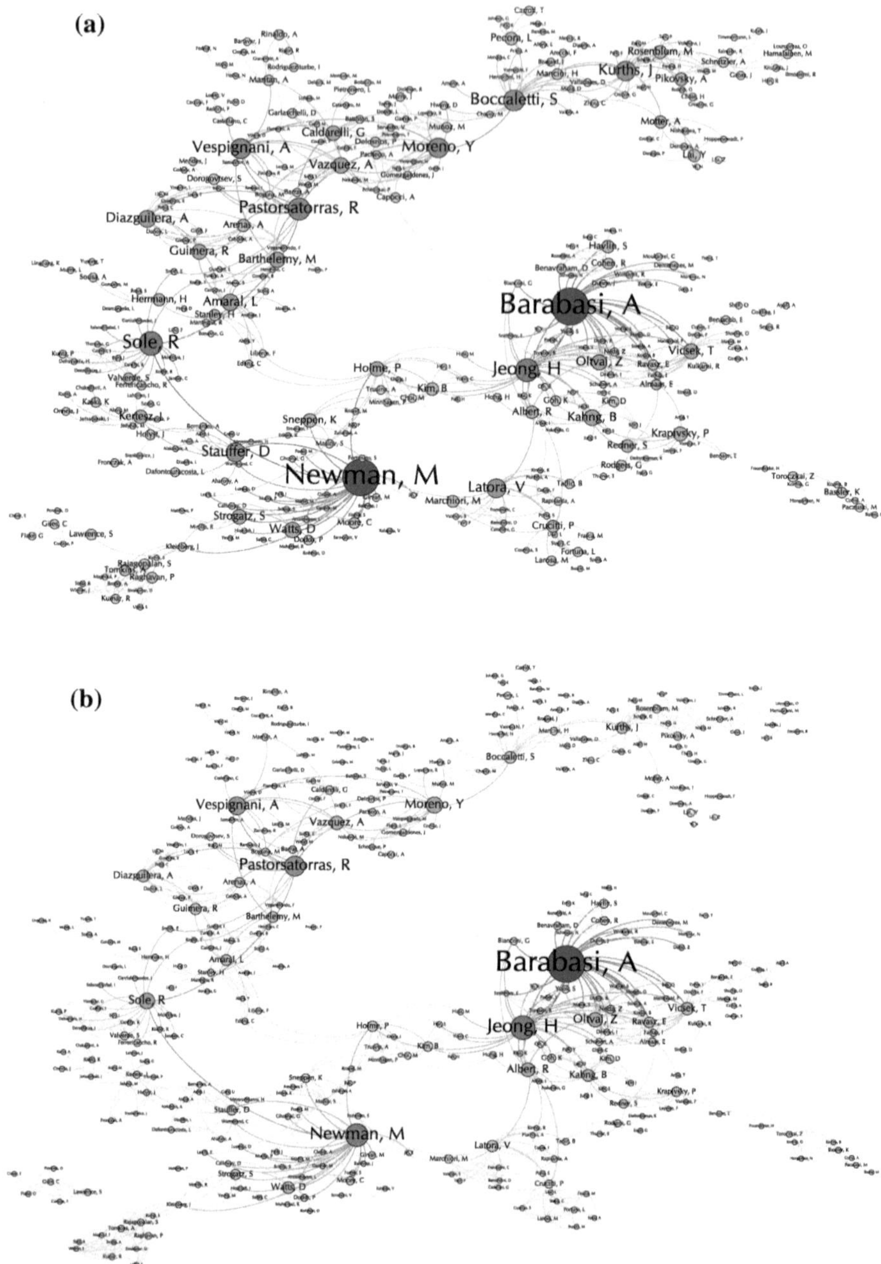

Fig. 10.1 Co–authorship network scientists. **a** PageRank. **b** Exponential ranking $\mu = 0.1$

Table 10.2 Top 10 rankings

	PageRank	Exp Rank μ = 0.1	Exp Rank μ = 1
1	Barabási, A	Barabási, A	Barabási, A
2	Newman, M	Jeong, H	Newman, M
3	Sole, R	Newman, M	Jeong, H
4	Jeong, H	Pastorsatorras, R	Pastorsatorras, R
5	Pastorsatorras, R	Vespignani, A	Vespignani, A
6	Boccaletti, S	Moreno, Y	Moreno, Y
7	Vespignani, A	Sole, R	Sole, R
8	Moreno, Y	Oltvai, Z	Boccaletti, S
9	Kurths, J	Albert, R	Vazquez, A
10	Stauffer, D	Vazquez, A	Diazguilera, A

10.3 Convergence and Uniqueness

More formally, let us define the map $V : S^n \to S^n$, which maps

$$V(p) = \frac{\exp \frac{1}{\mu} A^\top p}{\| \exp \frac{1}{\mu} A^\top p \|_1}, \tag{10.13}$$

where $S^n = \{y \in \mathbb{R}^n_+ : \|y\|_1 = 1\}$, the n-dimensional unit simplex. For the proof of convergence we rely on mixed matrix norms , or subordinate norms, which are defined as

$$\|A\|_{p,q} = \max_{\|x\|_q = 1} \|Ax\|_p. \tag{10.14}$$

Denoting by $\|A\|_{\max} = \max_{ij} |A_{ij}|$, we have the following useful inequality

$$\|Ax\|_\infty = \max_i \|e_i^\top Ax\| \leq \|A\|_{\max} \cdot \|x\|_1,$$

hence

$$\|A\|_{\infty,1} \leq \|A\|_{\max} \tag{10.15}$$

where e_i is the i-th coordinate vector. Let us now take a look at the Jacobian of V, which can be expressed as

$$\frac{\partial V(p)_i}{\partial p_j} = \frac{\exp(\frac{1}{\mu}A^\top p)_i \frac{1}{\mu}A_{ji}}{\sum_l \exp(\frac{1}{\mu}A^\top p)_l} - \frac{\exp(\frac{1}{\mu}A^\top p)_i \sum_l \exp(\frac{1}{\mu}A^\top p)_l \frac{1}{\mu}A_{jl}}{\left(\sum_l \exp(\frac{1}{\mu}A^\top p)_l\right)^2}.$$

Now let $u = \exp(\frac{1}{\mu}A^\top p)$, and $q = \|u\|_1$. Then $V(p) = u/q$, and $\frac{\partial V(p)_i}{\partial p_j}$ can be simplified to

$$\frac{\partial V(p)_i}{\partial p_j} = \frac{1}{\mu}\left(\frac{u_i}{q}A_{ji} - \frac{1}{q^2}\sum_l u_i u_l A_{jl}\right)$$

or in matrix notation

$$V'(p) = \frac{1}{\mu}\left(\frac{1}{q}\operatorname{diag}(u) - \frac{1}{q^2}uu^\top\right)A^\top \tag{10.16}$$

at which point the following lemma is useful.

Lemma 10.1 *Denote by $M(p)$ the matrix $M(p) = \operatorname{diag}(p) - pp^\top$ where $p \in S^n$, then $\|M(p)\|_{1,\infty} \le 1$.*

Proof Note that $\|M(p)x\|_1 = \sum_{i=1}^n p_i|x_i - p^\top x|$. We need to find the maximum of this function on the unit box (that is, where $\|x\|_\infty = 1$). By convexity of norms, the maximum of $\|M(p)x\|_1$ is attained at the boundary, i.e. some vector $\sigma \in R^n$ with coordinates ± 1. Denoting by $I_+ = \{i : \sigma_i = 1\}$ the set of positive entries, and by $S_1 = \sum_{i \in I_+} p_i$ and $S_2 = 1 - S_1$. Then $p^\top \sigma = S_1 - S_2$, and we have

$$\begin{aligned}\|M(p)\sigma\|_1 &= \sum_{i=1}^n p_i|\sigma_i - S_1 + S_2| = \sum_{i \in I_+} p_i|1 - S_1 + S_2| + \sum_{i \notin I_+} p_i|1 + S_1 - S_2| \\ &= S_1(1 - S_1 + S_2) + S_2(1 + S_1 - S_2) = 1 - (S_1 - S_2)^2.\end{aligned}$$

Since $(S_1 - S_2)^2 \ge 0$, $\|M(p)\sigma\|_1 \le 1$. □

This immediately leads to the following proof that the map V converges.

Theorem 10.2 *For $\mu > \frac{1}{2}(\max_{ij} A_{ij} - \min_{ij} A_{ij})$ the map V has a unique fixed point $p \in S^n$.*

Proof By the Banach fixed point theorem, this map has a unique fixed point if it is contractive. That is, there should be a $c < 1$ such that

$$\frac{\|V(p) - V(u)\|_1}{\|p - u\|_1} \le c, \tag{10.17}$$

for $p, u \in S^n$. That is, we should have $\|V'(p)\|_{1,1} \le c$. Since we can write $V'(p) = \frac{1}{\mu}M(V(p))A$, using the lemma and Eq. (10.15) we arrive at

$$\|V'(p)\|_{1,1} = \frac{1}{\mu}\|M(V(p))A\|_{1,1} \leq \frac{1}{\mu}\|M(V(p))\|_{1,\infty}\|A\|_{\infty,1} \leq \frac{1}{\mu}\|A\|_{\max}.$$

Since adding a constant to our matrix A does not change the vector $V(p)$, we can subtract $\frac{1}{2}(\min_{ij} A_{ij} + \max_{ij} A_{ij})$, and arrive at

$$\|V'(p)\|_{1,1} \leq \frac{1}{2\mu}(\max_{ij} A_{ij} - \min_{ij} A_{ij}).$$

Hence, if

$$\mu > \frac{1}{2}(\max_{ij} A_{ij} - \min_{ij} A_{ij}),$$

the map V is contractive and by the Banach fixed point theorem, it will have a unique fixed point, and iterates will converge to that point. □

For this lower bound on μ, we can guarantee convergence of the iteration. Below this lower bound, we choose nodes with more and more certainty. As we said in Sect. 10.2, when $\mu \to 0$ the probabilities $p_i = 1/|M|$ for i in some set M of nodes with maximal reputation r_i. In the iteration this means only nodes with the highest reputation can become judges. Since we completely trust his judgements, whatever node(s) he assigns the highest reputation will be the next judge. Unless everyone always agrees on the node with the highest reputation, cycles of judges pointing to the next judge will emerge.

For example, if we take $\mu \to 0$ for the example network given in Table 10.1, we cycle as follows. We start out with $p(0) = 1/n$, and the average reputation will be highest for nodes a and c, and they will be chosen as judge with probability $1/2$. In the next iteration the average reputation will be $1/2$ for nodes a, b and c and zero for d and e. Hence, one of the nodes a, b and c will be selected as judge, and the average reputation is $2/3$ for a and c, and $1/3$ for b. Now we are back where we were after the first iteration, since a and c both have the same maximal reputation, and they are chosen as judge each with probability $1/2$, as summarized in Table 10.1c.

References

1. Abrams, Z, McGrew, R and Plotkin, S (2004). Keeping Peers Honest in EigenTrust. In 2nd Workshop on the Economics of Peer-to-Peer Systems.
2. Altafini C (2012) Dynamics of opinion forming in structurally balanced social networks. PloS one 7(6):e38135. doi:10.1371/journal.pone.0038135
3. Anderson SP, de de Palma, A and Thisse, JF, (1992) Discrete Choice Theory of Product Differentiation. The MIT Press, Cambridge
4. Bonacich, P (1987). Power and centrality: A family of measures. American journal of sociology.
5. Bonacich P (2007) Some unique properties of eigenvector centrality. Social Networks 29:555–564. doi:10.1016/j.socnet.2007.04.002
6. Brin S, Page L (1998) The anatomy of a large-scale hypertextual Web search engine. Computer Networks and ISDN Systems 30(1–7):107–117. doi:10.1016/S0169-7552(98)00110-X

7. Christley RM, Pinchbeck GL, Bowers RG, Clancy D, French NP et al (2005) Infection in social networks: using network analysis to identify high-risk individuals. American journal of epidemiology 162(10):1024–31. doi:10.1093/aje/kwi308
8. De P, Singh AE, Wong T, Yacoub W, Jolly AM (2004) Sexual network analysis of a gonorrhoea outbreak. Sexually transmitted infections 80(4):280–5. doi:10.1136/sti.2003.007187
9. De Kerchove, C and Van Dooren, P (2008). The PageTrust algorithm: how to rank web pages when negative links are allowed. In Proceedings of the SIAM International Conference on Data Mining, pages 346–352. SIAM.
10. DeGroot MH (1974) Reaching a consensus. Journal of the American Statistical Association 69(345):118–121. doi:10.2307/2285509
11. Freeman LC (1977) A Set of Measures of Centrality Based on Betweenness. Sociometry 40(1):35. doi:10.2307/3033543
12. Freeman LC (1978) Centrality in social networks conceptual clarification. Social Networks 1(3):215–239. doi:10.1016/0378-8733(78)90021-7
13. Guha, R, Kumar, R, Raghavan, P and Tomkins, A (2004). Propagation of trust and distrust. In Proceedings of the 13th conference on World Wide Web - WWW '04, page 403. ACM Press, New York, New York, USA. ISBN 158113844X. doi:10.1145/988672.988727
14. Kamvar, SD, Schlosser, MT and Garcia-Molina, H (2003). The Eigentrust algorithm for reputation management in P2P networks. In Proceedings of the twelfth international conference on World Wide Web - WWW '03, page 640. ACM Press, New York, New York, USA. ISBN 1581136803. doi:10.1145/775152.775242
15. Kleinberg JM (1999) Authoritative Sources in a Hyperlinked Environment. Journal of the ACM 46(5):604–632. doi:10.1145/324133.324140
16. Maoz Z, Terris LG, Kuperman RD, Talmud I (2008) What Is the Enemy of My Enemy? Causes and Consequences of Imbalanced International Relations, 1816–2001. The Journal of Politics 69(01):100–115. doi:10.1111/j.1468-2508.2007.00497.x
17. Massa, P and Avesani, P (2005). Controversial Users demand Local Trust Metrics : an Experimental Study on Epinions. In Proceedings of the National Conference on Artificial Intelligence, pages 121–126.
18. Massa, P and Hayes, C (2005). Page-reRank: Using Trusted Links to Re-Rank Authority. In The 2005 IEEE/WIC/ACM International Conference on Web Intelligence (WI'05), pages 614–617. IEEE. ISBN 0-7695-2415-X. doi:10.1109/WI.2005.112
19. Newman, MEJ (2006). Finding community structure in networks using the eigenvectors of matrices. Physical Review E, 74(3):036104+. doi:10.1103/PhysRevE.74.036104
20. Olfati-Saber R, Fax JA, Murray RM (2007) Consensus and cooperation in networked multi-agent systems. Proceedings of the IEEE 95(1):215–233
21. Perra N, Fortunato S (2008) Spectral centrality measures in complex networks. Physical Review E 78(3):036107. doi:10.1103/PhysRevE.78.036107
22. Resnick P, Zeckhauser R, Swanson J, Lockwood K (2006) The value of reputation on eBay: A controlled experiment. Experimental Economics 9(2):79–101. doi:10.1007/s10683-006-4309-2
23. Shi, G, Proutiere, A, Johansson, M, Baras, JS and Johansson, KH (2013). The evolution of beliefs over signed social networks. [arXiv]1307.0539.
24. Szell M, Lambiotte R, Thurner S (2010) Multirelational organization of large-scale social networks in an online world. Proceedings of the National Academy of Sciences of the United States of America 107(31):13636–41. doi:10.1073/pnas.1004008107. [arXiv]1003.5137
25. Watts DJ (2002) A simple model of global cascades on random networks. Management 99(9):

Chapter 11
Conclusion

In this thesis we have explored two broad subjects: community detection and negative links. The latter subject is however also related to community detection, since networks with negative links are often believed to be organized into factions, such that positive links fall within factions and negative links in between them. We have seen how we can address the issue of the resolution limit, and suggested a very simple model (CPM) that circumvents this problem. In addition, CPM has a very natural interpretation: each community is expected to have a density of at least γ_{CPM}, while the density between two communities should be less then γ_{CPM}. Choosing some particular γ_{CPM} is not straightforward however and depends on the network in question. Nevertheless we were able to provide some insight into the different partitions returned for some γ_{CPM}. In particular, we introduced the notion of the "significance" of a partition, which helps in choosing some meaningful resolution parameter γ_{CPM}.

It is in some sense ironic that we return to the significance of a partition. In first instance, the popular method of modularity [3] was introduced in order to choose some "significant" level in an hierarchical clustering method. Because this method suffered from a resolution limit, we introduced the Constant Potts Model (CPM) that didn't rely on any comparison to a random graph. Yet, in order to determine a meaningful resolution, we returned to some comparison to a random graph. In this sense, we are back at square one: we have some single measure in order to determine some "significant" level.

This makes one wonder whether there exists any method that is capable of always detecting the "correct" partition. As we have seen, the problem of the resolution limit is usually associated to depending on some graph properties beyond the immediate link—only local methods do not seem to suffer from the resolution limit. Yet, a local method cannot be used to decide whether a partition is "meaningful" or not. In this sense, we might conjecture, in similar spirit as [2] his "impossibility theorem on clustering", that no community detection method exists that is both scale invariant and, in some vague notion, "meaningful".

Concerning negative links and social balance, we have shown that only the model $\dot{X} = XX^{\top}$ attains social balance generically. This implies that for almost any initial

V. Traag, *Algorithms and Dynamical Models for Communities and Reputation in Social Networks*, Springer Theses, DOI: 10.1007/978-3-319-06391-1_11,

condition, this model will converge to social balance. Moreover, once some network has attained social balance, for almost all perturbations away from social balance, the dynamics will return to social balance. This explains why we see so often networks split in two opposing camps.

In addition, the model $\dot{X} = XX^\top$ seems to be able to explain the evolution of cooperation through indirect reciprocity if reputations are private. It had been theorized that humans developed language so they could gossip about others, in order to strengthen their social network and sustain larger group sizes [1]. Yet our analysis suggests a subtly different mechanism: gossip didn't evolve to strengthen social networks but to maintain cooperation and dispel defectors. It is therefore ironic that the model predicts a split in two factions: even though gossip might have evolved to keep larger groups together, as a by product it seems to split groups in two. Whereas gossip was argued to be inclusive (it would integrate members of some social group), it also is exclusive (it repels members from different groups).

Nonetheless, the models currently analysed do exhibit several unrealistic features, we would like to address: (1) an all-to-all topology; (2) dynamics that blow-up in finite time; and (3) homogeneity of all agents. Although most of these issues can be addressed by specifying different dynamics, the resulting models are much more difficult to analyse, thereby limiting our understanding. Although the two models are somewhat simple, they are also tractable, and what we lose in truthfulness, we gain in deeper insights: in simplicity lies progress. Our current analysis offers a quite complete understanding for these relatively simple models.

References

1. Dunbar RIM (1998) Grooming, gossip, and the evolution of language. Harvard University Press, Cambridge. ISBN 0674363361
2. Kleinberg J (2003) An impossibility theorem for clustering. In: Advances in neural information processing systems, MIT Press, Cambridge. ISBN 0-262-02550-7
3. Newman M, Girvan M (2004) Finding and evaluating community structure in networks. Physical Review E 69(2):026113. doi:10.1103/PhysRevE.69.026113

Biography of Author

Vincent Traag is a complex networks researcher, and is currently analysing elite networks of Indonesia as a postdoc associated to the KITLV. He obtained his Masters degree (Cum Laude) in sociology at the University of Amsterdam (the Netherlands) in 2008, but also had a background in computer science and mathematics. He tried to combine his expertise in mathematics and sociology by focusing on social networks, and decided to pursue a Ph.D. in applied mathematics at the Université catholique de Louvain (Belgium) under the supervision of Paul Van Dooren and Yurii Nesterov. In his dissertation, Traag covers a wide range of topics including community detection in complex networks and dynamics of social balance, and published in a wide variety of highly regarded journals. He successfully defended his thesis in 2013.

Publications Related to This Thesis

Bruggeman, J, Traag, VA and Uitermark, J (2012). Detecting Communities through Network Data. *American Sociological Review*, 77(6):1050–1063. doi: 10.1177/0003122412463574.

Csáji, B, Browet, A, Traag, VA, Delvenne, JC, Huens, E et al. (2012). Exploring mobility of mobile users. *Physica A*, 392(6):1459–1473. doi: 10.1016/j.physa.2012.11.040.

Lupu, Y and Traag, VA (2012). Trading Communities, the Networked Structure of International Relations, and the Kantian Peace. *Journal of Conflict Resolution*. doi: 10.1177/0022002712453708.

Traag, VA, Browet, A, Calabrese, F and Morlot, F (2011). Social Event Detection in Massive Mobile Phone Data Using Probabilistic Location Inference. In *Proceedings IEEE SocialCom'2011*, pages 625–628. IEEE. doi: 10.1109/PASSAT/SocialCom.2011.133.

Traag, VA and Bruggeman, J (2009). Community detection in networks with positive and negative links. *Physical Review E*, 80(3):036115. doi: 10.1103/PhysRevE.80.036115. arXiv:0811.2329.

V. Traag, *Algorithms and Dynamical Models for Communities and Reputation in Social Networks*, Springer Theses, DOI: 10.1007/978-3-319-06391-1,

Traag, VA, Krings, G and Van Dooren, P (2013). Significant scales in community structure. *submitted*. arXiv:1306.3398.

Traag, VA, Nesterov, Y and Van Dooren, P (2010). Exponential Ranking: taking into account negative links. *LNCS*, 6430:192–202. doi: 10.1007/978-3-642-16567-2.

Traag, VA, Van Dooren, P and De Leenheer, P (2013). Dynamical models explaining social balance and evolution of cooperation. *PLoS ONE*, 8(4):e60063. doi: 10.1371/journal.pone.0060063. arXiv:1207.6588.

Traag, VA, Van Dooren, P and Nesterov, Y (2011a). Indirect reciprocity through gossiping can lead to cooperative clusters. In *IEEE Symposium on Artificial Life 2011*, pages 154–161. IEEE. doi: 10.1109/ALIFE.2011.5954642.

Traag, VA, Van Dooren, P and Nesterov, Y (2011b). Narrow scope for resolution-limit-free community detection. *Physical Review E*, 84(1):016114. doi: 10.1103/PhysRevE.84.016114. arXiv:1104.3083.

Index

V. Traag, *Algorithms and Dynamical Models for Communities and Reputation in Social Networks*, Springer Theses, DOI: 10.1007/978-3-319-06391-1,